ANOXIA

Edited by **Pamela Padilla**

Anoxia
Edited by Pamela Padilla

Published by InTech
Janeza Trdine 9, 51000 Rijeka, Croatia

Publishing Process Manager Marina Jozipovic

Technical Editor Teodora Smiljanic

Cover Designer InTech Design Team

Additional hard copies can be obtained from orders@intechopen.com

Anoxia, Edited by Pamela Padilla
p. cm.
ISBN 978-953-307-664-5

For EU product safety concerns:
IN TECH d.o.o., Prolaz Marije Krucifikse Kozulić 3, 51000 Rijeka, Croatia,
info@intechopen.com or visit our website at intechopen.com.

Meet the editor

Dr Pamela Padilla is an Associate Professor in the Department of Biological Sciences at the University of North Texas. She received her PhD from the University of New Mexico, conducted her post-doctoral research at the Fred Hutchinson Cancer Research Center and was a Glenn Visiting Scholar at the Salk Institute. She has received funding support from the National Institutes of Health and National Science Foundation. She received prestigious awards and honors including being named a Kavli Fellow and awarded an Early Career Award for Research and Creativity at UNT and an NSF CAREER. Her research interests include environmentally induced arrest in various organisms (nematode Caenorhabditis elegans and killifish Austrofundulus limnaeus). Currently, her laboratory focus involves understanding how genetics and environmental factors, such as diet, influence the response to oxygen deprivation. At UNT she teaches undergraduate and graduate courses in the field of Genetics and Stress Physiology.

Contents

Preface XI

Part 1 Model Systems 1

Chapter 1 **Anoxia Tolerance During Vertebrate Development - Insights from Studies on the Annual Killifish *Austrofundulus limnaeus* 3**
Jason E. Podrabsky, Claire L. Riggs and Jeffrey M. Duerr

Chapter 2 **Anoxia-Induced Suspended Animation in *Caenorhabditis elegans* 25**
Pamela A. Padilla, Jo M. Goy and Vinita A. Hajeri

Part 2 Human Health Related Issues 59

Chapter 3 **Resistance of Multipotent Mesenchymal Stromal Cells to Anoxia *In Vitro* 61**
L.B. Buravkova, E.R. Andreeva, J.V. Rylova and A.I. Grigoriev

Chapter 4 **The Protective Role of Hypoxic Preconditioning in CNS 81**
Li-Ying Wu, Ling-Ling Zhu and Ming Fan

Part 3 Ecological Systems 97

Chapter 5 **Increase in Anoxia in Lake Victoria and Its Effects on the Fishery 99**
Murithi Njiru, Chrisphine Nyamweya, John Gichuki, Rose Mugidde, Oliva Mkumbo and Frans Witte

Part 4 Plant Methodology 129

Chapter 6 **Connection Between Structural Changes and Electrical Parameters of Pea Root Tissue Under Anoxia 131**
Eszter Vozáry, Ildikó Jócsák, Magdolna Droppa and Károly Bóka

Preface

Oxygen is essential to biological systems; life as we know it would not exist if this important molecule was not a component of our Earth. During evolution, organisms have adapted to ever-changing oxygen levels within their environment. Even organisms such as humans, which are very sensitive to oxygen deprivation, respond to hypoxia and anoxia at the molecular, cellular and organ level. Other organisms, such as the killifish embryo (*Austrofundulus limnaeus*) are very tolerant to anoxia and the mechanisms leading to anoxia or hypoxia tolerance are being investigated. Some of the responses to oxygen decrease, such as HIF-1 signaling, are evolutionarily conserved among metazoans, while other responses have only been observed in a few organisms.

Although the book is not large in terms of number of chapters, and its overview of oxygen deprivation response and tolerance is not completely comprehensive, it does have an interesting assorted range of topics related to severe oxygen deprivation (hypoxia and anoxia). Individual chapters will be of interest to those within the specific fields, whereas the book as a whole is a good resource to obtain an overview of research areas in the ever-increasing field of oxygen deprivation. The contributing authors are from many areas of the world; 11 academic institutes in seven different countries on four continents are represented in this book. The diversity in authorship underscores the idea that individuals throughout the world are thinking of and investigating oxygen deprivation affects on living systems.

The first two chapters are devoted to model systems used to study hypometabolic states and anoxia tolerance. In the first chapter anoxia tolerance in the annual killifish *Austrofundulus limnaeus* is evaluated. Annual killifish have evolved towards obtaining the capacity to survive in ephemeral ponds in regions of South America and Africa, where wet and dry weather interchanges within a season. What happens in the ponds as a result of these seasonal changes, during the dry season, the embryos can become buried in the sediment and exposed to severe oxygen deprivation. Under favorable conditions these remarkable embryos have the ability to enter into an obligate diapause (diapause II), which is an arrested, hypometabolic state that confers anoxia resistance. Chapter 1 reviews the physiological and metabolic changes associated with anoxia survival in killifish embryos. The second chapter focuses on the genetic and cellular response *Caenorhabditis elegans* has to anoxia. *C. elegans*, which is a well-known model system, can survive anoxia by entering a reversible, hypometabolic state referred to as suspended animation. Chapter 2 highlights the cellular or anatomical changes embryos or adult animals have to anoxia exposure. Furthermore, genes that are required for anoxia survival are described. The use of model systems, *A. limnaeus* and *C. elegans*, to investigate anoxic tolerance could be used for comparative analysis between organisms that are tolerant and sensitive to anoxia.

The third and fourth chapters are devoted to understanding oxygen deprivation in models that are certainly relevant to human health questions. The third chapter highlights how oxygen balance is important for tissue physiology and cellular function. In this case, the resistance to anoxia in multipotent mesenchymal stromal cells (MMSCs) is discussed. An important finding is that oxygen is a crucial component determining stem cell homeostasis within the tissue niche. Low oxygen tension is an important regulator of MMSCs maintenance. Given the importance of understanding stem cell biology, the effect of microenvironment and oxygen tension is of interest. The fourth chapter reviews hypoxic preconditioning and how it protects against anoxia exposure in a system that is extremely sensitive to oxygen deprivation, the central nervous system. Additionally, the signaling pathways that have a role in hypoxia preconditioning (Ex: HIF-1 signaling) are discussed in the context of hypoxia precondition. Understanding the mechanisms that underlie hypoxia preconditioning will undoubtedly lead to a deeper understanding of anoxia response and survival.

The last two chapters highlight the important fact that hypoxia can affect ecosystems, such as lakes, or agriculturally important systems such as the pea plant. The fifth chapter focuses on the impact hypoxia has on Lake Victoria fishery whereas the sixth chapter focuses on the idea that structural changes within plants exposed to hypoxia can be detected using electrical impedance spectroscopy. Lake Victoria, which is the second largest freshwater lake in the world and the largest in the tropics, is an important biological and economic resource. However, all too often environmental changes, caused by human activities, increase the occurrence of hypoxia in large bodies of water. This chapter discusses characteristics of Lake Victoria and the impact hypoxia has on fish at the anatomical, physiological and population level. Given the increasing world population and anthropogenic factors that influence ecological systems, studies such as those conducted at Lake Victoria are of interest. Last, but not least, the sixth chapter provides evidence that electrical impedance measurements are a non-invasive method to follow changes within pea seedlings' tissue. Tissue changes within plants are often induced by severe oxygen deprivation caused by flooding. Because these changes can be detected on young seedlings, visible symptoms induced by oxygen deprivation stress can be observed. Such an approach can be valuable to better understand the physiology of flood-tolerant and flood-resistant agriculture plants.

I wish to express my special thanks to all the authors who contributed their excellent articles to this book. Furthermore, I wish to express my gratitude to Marina Jozipovic, Publishing Process Manager, for editorial support.

Pamela A. Padilla,
Department of Biological Sciences
University of North Texas
Denton TX,
USA

Part 1

Model Systems

1

Anoxia Tolerance During Vertebrate Development - Insights from Studies on the Annual Killifish *Austrofundulus limnaeus*

Jason E. Podrabsky[1], Claire L. Riggs[1] and Jeffrey M. Duerr[2]
[1]*Portland State University,*
[2]*George Fox University,*
USA

1. Introduction

With a rare few exceptions, vertebrates are extremely sensitive to a lack of oxygen and can survive for only brief episodes of oxygen deprivation (Nilsson and Lutz, 2004). In fact, when differences in body temperature are taken into account, endotherm and ectotherm vertebrates share a very similar survival time in anoxia, and similar symptoms of cellular and tissue damage and death (Nilsson and Lutz, 2004). Thus, there appears to be a common limit to survival of anoxia in most vertebrates that may be supported by common limits to metabolic and physiologic systems. The few exceptions to this rule (all aquatic vertebrates) have developed novel mechanisms to support tolerance of long-term anoxia (Nilsson and Lutz, 2004; Podrabsky et al., 2007).

It has long been appreciated that fetal and neonatal mammals can tolerate much longer exposures to anoxia when compared to their adult counterparts (e.g. Kabat, 1940; Fazekas et al., 1941; Glass et al., 1944; Adolph, 1969). In fact, for the great majority of vertebrates, tolerance of anoxia is highest in the earliest developmental stages and is progressively lost during development (Fig. 1). This relationship is surprisingly consistent among a wide diversity of lineages and reproductive strategies with a correlation coefficient of -0.87 when data for all "typical" vertebrates are included. Embryos of the annual killifish *Austrofundulus limnaeus* clearly exhibit a different level (2 orders of magnitude greater) and pattern (increase in anoxia tolerance during early development) of anoxia tolerance compared to the other vertebrates for which data exist (Fig. 1). Embryos of *A. limnaeus* gain and then lose the ability to survive prolonged bouts of anoxia as a normal part of their developmental program (Podrabsky et al., 2007; Fig. 1). This unique life history pattern allows for the biological mechanisms that support tolerance of anoxia to be studied in a comparative context within a single species. In addition, these embryos respond to anoxia using the same basic metabolic pathways used by more typical vertebrates, and yet they can survive for months without oxygen. This fact suggests that novel mechanisms of anoxia tolerance *that have not been explored in other systems* may be operating or be induced in this species. Thus, a deeper understanding of the mechanisms that underlie tolerance of anoxia in this species has the potential to transform our understanding of anoxia tolerance, and may lead to new avenues for minimizing or preventing damage due to restriction of oxygen during vertebrate development, and as a consequence of heart attack or stroke.

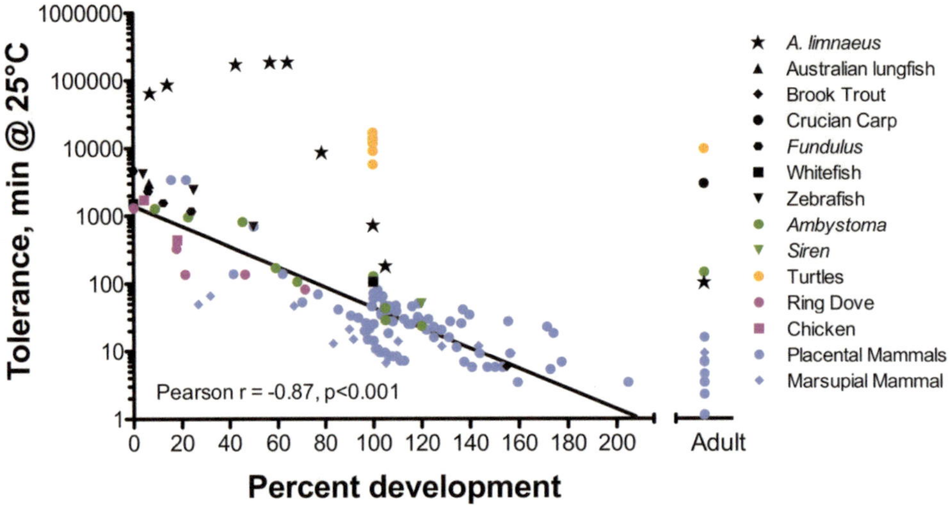

Fig. 1. Tolerance of anoxia declines during normal development in a wide variety of vertebrate embryos. Embryos of *Austrofundulus limnaeus* are unique among vertebrates in their extreme tolerance of anoxia beyond that predicted as a consequence of developmental stage. Survival values have been corrected for body or experimental temperature to a common temperature of 25°C assuming a Q_{10} of 2. The regression line was calculated based on all species and developmental stages excluding the two obvious outliers, embryos of the annual killifish *A. limnaeus* and turtle hatchlings. 100% of development was set at birth or hatching for turtles, birds and placental mammals, emergence from the pouch for the Virginia opossum, and at completion of larval development for fish and amphibians. Data were taken from: *A. limnaeus* adults and larvae, Podrabsky (new data) and embryos, Podrabsky et al., 2007; Australian lungfish (*Neoceratodus forsteri*), Mueller et al., 2011; brook trout (*Salvelinus fontinalis*), Shepard, 1955; crucian carp (*Carassius carassius*) and adult turtles (*Trachemys scripta, Chrysemys picta*), Nilsson and Lutz, 2004; *Fundulus sp.*, Loeb, 1894; whitefish (*Coregonus clupeiformis*), Hall, 1925; zebrafish (*Danio rerio*), Mendelsohn et al., 2008; salamanders (*Ambystoma sp.*), Rose et al., 1971, Weigmann and Altig, 1975, Adolph, 1979; *Siren intermedia*, Weigmann and Altig, 1975; turtle hatchlings (*Graptemys geographica, Chelydra serpentine, Trachemys scripta, Chrysemys picta, Emydoidea blandingii, Malaclemys terrapin, Terrapene ornate*), Dinkelacker et al., 2005; ring dove (*Streptopelia risoria*) Riddle, 1924; chicken (*Gallus gallus*), Nelson, 1958 and Di Carlo and Litovitz, 1999; dog (*Canis lupus familiaris*), Kabat, 1940; dog, cat (*Felis catus*), ground squirrel (*Citellus tridecemlineatus*), rat (*Rattus norvegicus*), golden hamster (*Mesocricetus auratus*), Adolph, 1969; guinea pig (*Cavia porcellus*) and rabbit (*Oryctolagus cuniculus*), Glass et al., 1944; sheep (*Ovis aries*), Dawes et al., 1959; rabbit blastocysts, Daniel, 1968; rhesus macaque (*Macaca mulatta*), Dawes et al., 1960; mouse (*Mus musculus*), Ingalls et al., 1950; Virginia opossum (*Didelphis virginiana*), Rinka and Miller, 1967.

2. The life history of *Austrofundulus limnaeus*

Annual killifish are a group of teleost fish that have evolved to survive in ephemeral ponds in regions of Africa and South America that experience pronounced dry and rainy seasons. Populations persist in isolated ponds due to production of drought-tolerant embryos that can enter embryonic diapause (Wourms, 1972a, 1972b; Podrabsky et al., 2010b). Embryos may be exposed to a variety of environmental extremes during their normal development including severe hypoxia or anoxia from being buried in the pond sediments (Podrabsky et al., 1998). Development may be arrested at any of three distinct stages of embryonic diapause in this group of fish (diapause I, II and III), although not all species arrest in all three stages (Wourms, 1972b). In our stock of *A. limnaeus*, embryos routinely arrest in the laboratory at diapause II and III (Podrabsky and Hand, 1999). Although most embryos arrest in diapause II, some may bypass this stage of diapause and develop directly to diapause III (Wourms, 1972b; Podrabsky et al., 2010a). These embryos are called "escape embryos" (Wourms, 1972b) and very little is known about the physiology of their anoxia tolerance, although they appear to have a different metabolic poise when compared to embryos that enter diapause II (Chennault and Podrabsky, 2010).

Associated with early development and entry into diapause II is the acquisition of extreme tolerance of anoxia. Diapause II embryos have a lethal time to 50% mortality (LT_{50}) of about 65 days of anoxia at 25°C and some embryos can survive for over 120 days of complete anoxia (Podrabsky et al., 2007). Diapause II embryos arrest midway through development at about 25 days post-fertilization (dpf) at 25°C. At this point in development, Wourms' stage (WS) 32, they possess the foundations of the central nervous system, including clearly defined fore-, mid-, and hindbrain regions, optic cups, otic vesicles, 38 pairs of somites, and a beating tubular heart (Wourms, 1972a, 1972b; Podrabsky and Hand, 1999). Extreme tolerance of anoxia is retained for at least 4 days of post-diapause II (dpd) development in WS 36 embryos that have much higher metabolic activity than diapause II embryos, and have experienced significant growth and differentiation in the brain, and circulatory systems (Podrabsky and Hand, 1999; Podrabsky et al., 2007). As the embryos continue to develop beyond WS 36, they lose their exceptional anoxia tolerance, and by the time embryonic development is complete they can survive for less than 24 hrs. This gain and then loss of anoxia tolerance makes *A. limnaeus* a unique and powerful model for investigation of the mechanisms that support long-term tolerance of anoxia in vertebrates.

3. Metabolic rate depression and survival of anoxia

The ability to reversibly enter a state of metabolic depression and to coordinately down-regulate ATP production and consumption are considered hallmarks of anoxia tolerant vertebrates (Hand and Hardewig, 1996; Hand, 1998; Krumschnabel, 2000; Hochachka and Somero, 2002). In embryos of *A. limnaeus* anoxia induces a complete cessation of development (Podrabsky et al., 2007). Embryos in diapause II may be pre-adapted for survival of anoxia because they are already in a state of profound metabolic depression (Podrabsky and Hand, 1999). However, post-diapause II embryos are actively developing with a metabolic rate an order of magnitude higher than embryos in diapause II (Podrabsky and Hand, 1999). Embryos at 4 dpd (WS 36, LT_{50} ~ 65 days of anoxia) experience a 94% reduction of heat flow within 16 hrs of exposure to anoxia (Fig. 2A; Podrabsky, Menze, and Hand, unpublished data). Associated with this reduction in heat flow is an over 80%

reduction of ATP content (Fig 2B; Podrabsky, Menze, and Hand, unpublished data). While the metabolic depression is not surprising, the large-scale loss of whole-embryo ATP is a striking result because it indicates a lack of coordination between ATP production and consumption. This is a major departure from other anoxia tolerant vertebrates, and indicates a possible novel cellular survival strategy. This is an important point, because it indicates that loss of cellular ATP does not have to lead to cell death in vertebrate cells, although it typically does in anoxia tolerant and sensitive species. Determining the cellular and molecular basis of survival in these cells despite the loss of ATP may be a promising avenue for developing novel treatments of anoxia in anoxia sensitive vertebrates such as mammals.

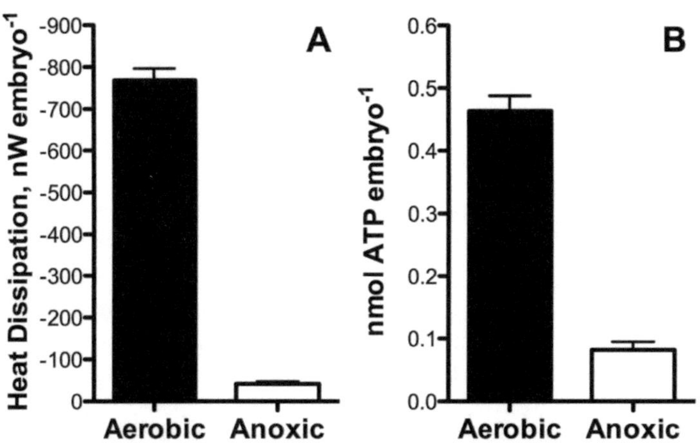

Fig. 2. Heat dissipation and ATP content of WS 36 embryos (4 dpd, LT_{50} ~65 days of anoxia) exposed to 14-16 hrs of anoxia. (A) Heat dissipation is reduced by 94% in anoxic compared to aerobic embryos. (B) Anoxia causes a decline in ATP content of over 80%. (Podrabsky, Menze, and Hand, unpublished data).

4. Heart rate during anoxia

Exposure to anoxia or ischemia typically causes a slowing of heart rate (bradycardia) in embryonic and fetal vertebrates. This is in stark contrast to the adult response of increased heart rate (tachycardia). In embryos of *A. limnaeus* with maximal anoxia tolerance (WS 36) heart activity ceases within 24 hrs of exposure to anoxia, while late embryos (WS 40) and larvae maintain a severe bradycardia until death (Fig. 3; Fergusson-Kolmes and Podrabsky, 2007). Thus, in this species extreme tolerance of anoxia is associated with the ability to reversibly pause cardiac contractility. The mechanisms that allow the cessation of heart activity, or conversely that prevent cessation of the heart in later developmental stages are currently unknown. It is possible that increased neural and endocrine control over heart function explains this difference, and the ontogeny of cardiac physiology during annual killifish development should be the focus of future studies. These data may help to inform studies of the effects of hypoxia and anoxia on the developing mammalian fetus. There is some controversy about the susceptibility of fetal hearts to hypoxic exposures, with those measuring cessation of heart activity finding fetal hearts more susceptible to hypoxia than adult hearts (Ostadal et al., 1999). However, it is possible that fetal hearts stop beating as an adaptive response to anoxia, which seems more consistent with the overall greater tolerance of anoxia in fetal and neonatal mammals as discussed above.

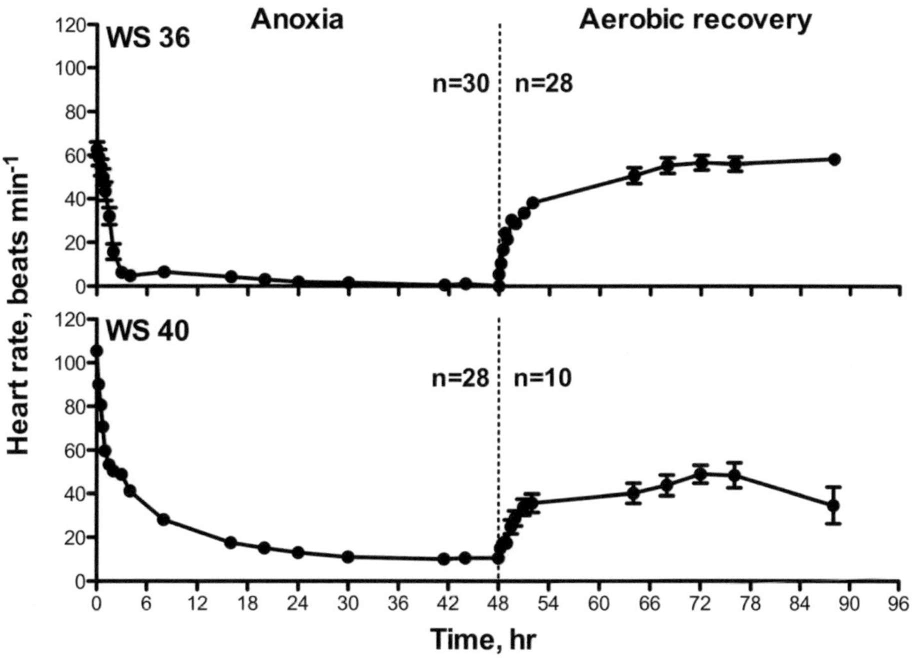

Fig. 3. Heart rate ceases in embryos with long-term tolerance of anoxia (WS 36), while embryos that have lost extreme tolerance of anoxia (WS 40) sustain a severe bradycardia in response to anoxia (Fergusson-Kolmes and Podrabsky, 2007). WS 36 (4 dpd), WS 40 (12 dpd).

5. Anaerobic metabolism and end-product accumulation

Anaerobic metabolism appears to be very constrained in vertebrates, with glycolytic production of lactate being the almost exclusive major end-product of anaerobic metabolism. The one major exception to this rule is the production of ethanol (in addition to a substantial amount of lactate) during exposure to anoxia in fish of the genus *Carassius* (goldfish and Crucian carp) and perhaps the bitterling (Shoubridge and Hochachka, 1980; Johnston and Bernard, 1983; van Waarde, 1991). However, to our knowledge no embryo or fetus has been reported to produce ethanol in response to anoxia. Embryos of *A. limnaeus* produce large quantities of lactate as their major anaerobic end-product (Fig. 4; Podrabsky et al., 2007). Thus, metabolic rate depression is absolutely key to the success of this survival strategy and there is a strong negative correlation between rate of lactate accumulation (a reasonable proxy for metabolic rate under anoxia) and survival times in anoxia (Fig. 5; Podrabsky et al., 2007). In addition, maximal survival times of anoxia in embryos of *A.*

limnaeus (just over 100 days) agree well with predicted exhaustion of embryonic glycogen stores (Podrabsky et al., 2007).

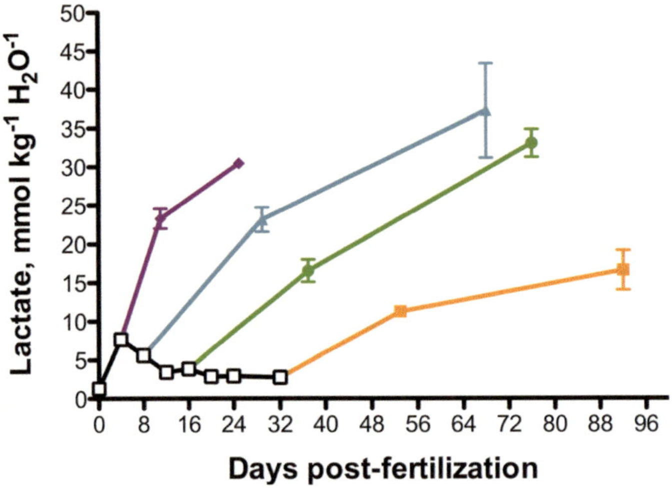

Fig. 4. Lactate is the major end-product of anaerobic metabolism in embryos of *A. limnaeus*. Open symbols represent normoxic values, while colored symbols represent accumulation of lactate under anoxia (Podrabsky et al., 2007).

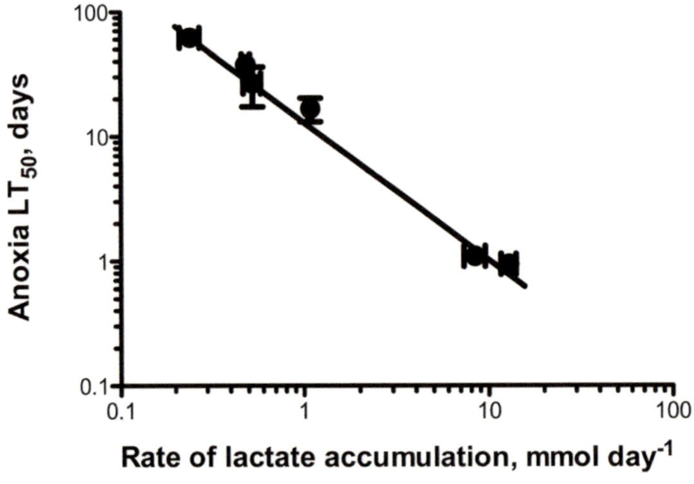

Fig. 5. The rate of lactate accumulation during exposure to anoxia is highly correlated with survival times in anoxia, supporting the importance of metabolic rate depression in survival of long-term anoxia (Podrabsky et al., 2007).

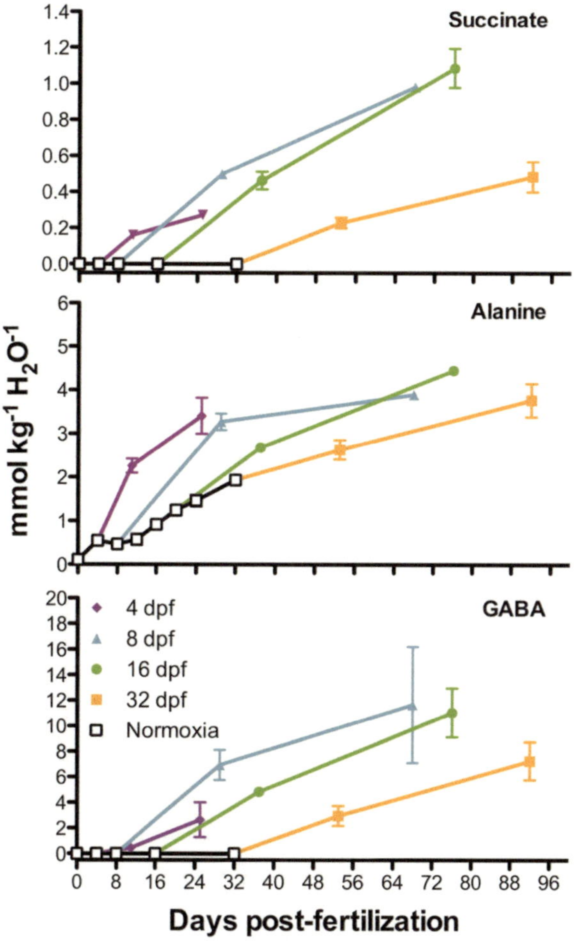

Fig. 6. Embryos of *A. limnaeus* accumulate succinate, alanine, and large quantities of γ-aminobutyrate (GABA) when exposed to anoxia. Open symbols represent normoxic values while colored symbols represent accumulation during exposure to anoxia (Podrabsky et al., 2007).

A. limnaeus embryos also accumulate substantial amounts of alanine, succinate, and γ-aminobutyrate (GABA) under anoxic conditions (Fig. 6; Podrabsky et al., 2007). Accumulation of succinate and alanine is common among invertebrate species that tolerate prolonged exposures to anoxia (Hochachka and Somero, 2002), and indicates that continued mitochondrial intermediary metabolism is essential to support anoxia tolerance in embryos of *A. limnaeus* (see below). Accumulation of GABA has been documented in the brains of adult turtles and crucian carp and has been implicated in the prevention of excitotoxic cell death through its role as an inhibitory neurotransmitter in the adult vertebrate nervous system (Nilsson and Lutz, 2004). However, the function of GABA as a neurotransmitter in the developing nervous system is complex. GABA has a generally excitatory affect on the developing nervous system in the few vertebrates in which it has been studied (Cherubini et al., 1991; Ben-Ari, 2002). In rats, GABA signaling is excitatory in late term fetuses and is only transiently transformed into an inhibitory neurotransmitter through the expression of specific chloride channels during the birthing process (Tyzio et al., 2006). Importantly, this change in neurotransmitter activity is associated with an increase in tolerance of oxygen deprivation that is thought to be important for survival of ischemia caused during labor. Levels of GABA are non-detectable in early embryos (through diapause II) of *A. limnaeus* during normoxic incubation (Podrabsky et al., 2007). When exposed to anoxia, GABA accumulates to high levels in embryos that possess extreme anoxia tolerance (Fig. 6; Podrabsky et al., 2007). These data suggest that GABA is not yet used as a neurotransmitter in these embryos, or that GABAergic neurons are in low abundance, despite the presence of a differentiated nervous system. In addition, the high levels of GABA accumulated (8-12 mM) may indicate a metabolic role for the production of GABA in embryos of *A. limnaeus*, in addition to or instead of a role as an inhibitory neurotransmitter. Further exploration of the role of GABA in survival of anoxia, and of the development of GABA signaling pathways in the developing nervous system are needed to resolve the role of accumulation of GABA during exposure to anoxia in developing vertebrates.

6. Mitochondrial physiology of developing embryos

Embryos of *A. limnaeus* develop an impressive tolerance of anoxia as part of their normal development, even under aerobic conditions. In fact, as early embryos develop and enter diapause II, they are consistently poised for anaerobic lactate production with over 50 times the capacity for lactate dehydrogenase (LDH) compared to citrate synthase (CS) activity (Fig. 7; Chennault and Podrabsky, 2010). Even during the highly aerobic regions of development in this species (see below) there is about 13 times more capacity for LDH activity compare to CS activity (Fig. 7). In fact, citrate synthase activity is low throughout most of embryonic development in this species, especially in those developmental stages that exhibit the greatest tolerance of anoxia. This low level of citrate synthase activity is consistent with what is known about mitochondrial physiology in this species (see below, Duerr and Podrabsky, 2010). Calorimetric:respirometric (C:R) ratios can be used as an indicator of the contribution of anaerobic metabolism to total heat production. C:R ratios indicate a significant contribution of anaerobic metabolism to total heat flow in early embryos through diapause II (Podrabsky and Hand, 1999). This anaerobic poise is lost in post-diapause II development when C:R ratios are indistinguishable from theoretical values based on a completely aerobic metabolism (Podrabsky and Hand, 1999). One strategy of these embryos may be to delay maturation of mitochondrial oxidative metabolism in order to prepare for exposures to anoxia in their natural environment. This may help to explain the slow rates of development in this species (about 38-40 dpf at 25°C to hatching if diapause II is not entered) compared to species with similar sized eggs and life histories such as *Fundulus heteroclitus* which complete development after 11-16 days at 20°C (DiMichele and Powers, 1984). There may be a trade-off between anoxia tolerance and developmental rate that is governed by the ability to retain the metabolic structure of very early embryos (cleavage and blastula stage) that relies on anaerobic rather than oxidative pathways to support development. The dependence of anoxia tolerant embryos on anaerobic metabolism even under aerobic conditions is supported by an extremely low activity of mitochondrial electron transport chain complexes and the almost complete absence of ATP synthase activity in mitochondria isolated from diapause II embryos (Fig. 8; Duerr and Podrabsky, 2010). In fact, the extremely low rates of oxygen consumption exhibited by mitochondria isolated from diapausing embryos can be attributed almost exclusively to proton leak (Duerr and Podrabsky, 2010). This apparent waste of metabolic fuel to maintain proton leak and therefore some small level of oxidative metabolism appears paradoxical.

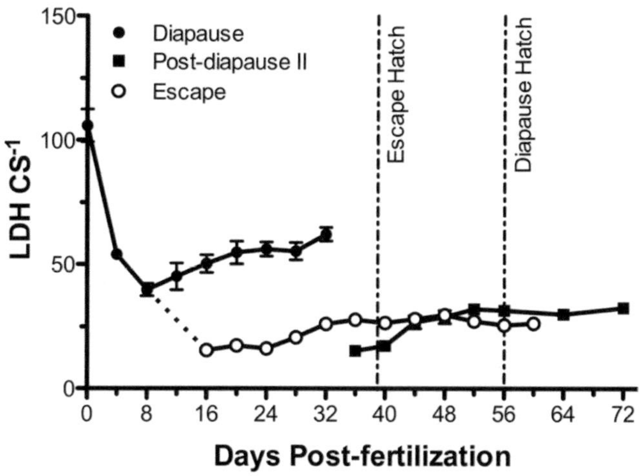

Fig. 7. Embryos of *A. limnaeus* have a far greater enzymatic capacity for lactate dehydrogenase activity compared to citrate synthase activity throughout development. It is especially high during early development and diapause II (Chennault and Podrabsky, 2010).

We hypothesize that mitochondria in the developing killifish are likely rather organized to support limited intermediary metabolism such as amino acid transamination, which appears to be important in the survival of *A. limnaeus* embryos exposed to anoxia (Podrabsky et al., 2007). For example, figure 8 illustrates that during diapause II and III, the ATP synthase (Complex V) is essentially inactive, though complexes II and IV are moderately active. Proton leak, as indicated previously, is also elevated during these same periods. One interpretation is that this is a succinate-dependent pathway for maintaining a limited proton-motive force. A nominal mitochondrial membrane potential is required for metabolite transporter function to provide substrate for the citric acid cycle and to export metabolites to the cytosol. We further hypothesize that the inner membrane of mitochondria from *A. limnaeus* embryos may exhibit elevated proton conductance as a mechanism to avoid extremely high membrane potentials and subsequent production of reactive oxygen species during transitions into and out of anoxia. This theory has yet to be tested experimentally. Mitochondrial densities within *A. limnaeus* embryos, as estimated by mtDNA content, increase by nearly four-fold as the embryos transition from diapause III to adult (Duerr and Podrabsky, unpublished data). These data clearly illustrate that the quantity and nature of mitochondria of *A. limnaeus* embryos are not consistent with maximal aerobic ATP production, but rather a role in supporting intermediary metabolism.

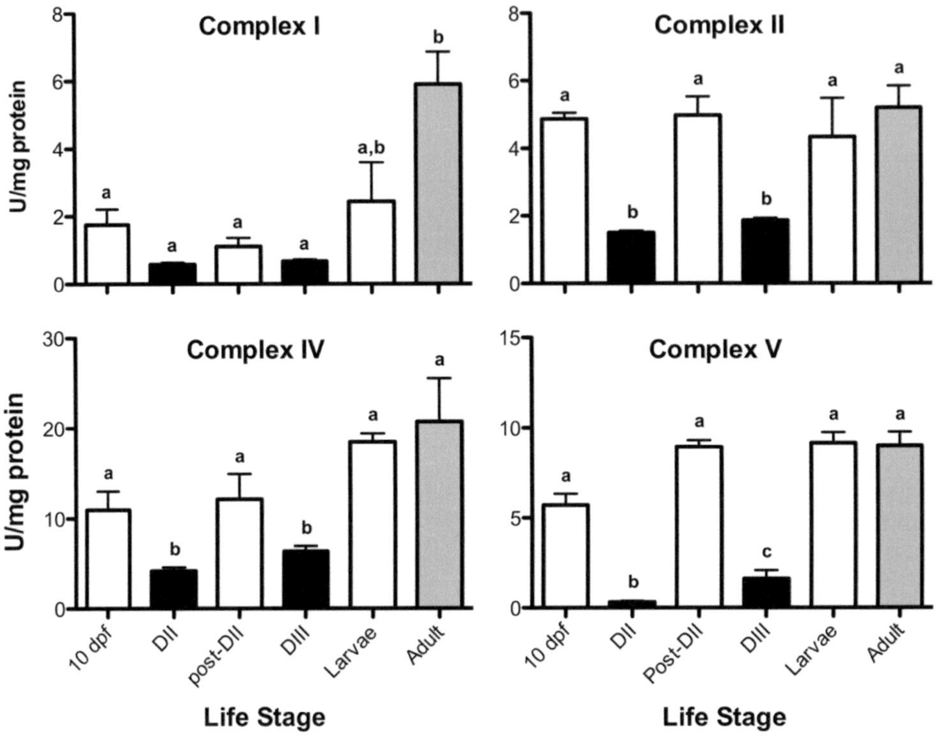

Fig. 8. Diapausing embryos have reduced activity of enzyme complexes involved in mitochondrial electron transport, including extremely low levels of ATP synthase (complex V) activity. DII = diapause II, DIII = diapause III, post-DII embryos are WS 39, larvae are 1-2 days post-hatch, adult data are for isolated liver mitochondria (Duerr and Podrabsky, 2010).

7. Preconditioning and anoxia tolerance

It has long been appreciated that short, non-lethal bouts of oxygen deprivation (ischemic or hypoxic) can induce a protective phenotype in mammalian tissues which supports reduced tissue damage and increased survival of subsequent more severe bouts of oxygen deprivation; a phenomenon known as ischemic or hypoxic preconditioning (Murray et al., 1986). Because of its clinical potential, preconditioning has been well-studied in mammalian systems, but few studies of preconditioning have been reported for other groups of vertebrates (Mulvey and Renshaw, 2000; Gamperl et al., 2001; Gamperl et al., 2004). A better understanding of the preconditioning-induced phenotype may lead to new therapies to reduce tissue damage and mortality as a result of heart attack or stroke. Preconditioning as defined in biomedical research on mammalian heart and brain tissue is very similar to induced tolerance as described by comparative physiologists. In both instances, an initial stressful event protects the organism or tissue in subsequent exposures. The advantages of this strategy are obvious, because in many situations (such as daily fluctuations in temperature, or tidal cycles) exposures to environmental stress are likely to occur repeatedly or intermittently. Preconditioning is very likely one manifestation of a highly conserved cellular stress response (Kültz, 2003, 2005). Thus, exploration of the cellular stress response in a variety of evolutionary lineages may shed light on the mechanisms of induced ischemia tolerance due to preconditioning.

7.1 Anoxic preconditioning

In embryos of *A. limnaeus*, survival of anoxia can be increased following an anoxic preconditioning (AP) regime of 24 hrs of anoxia followed by 24 hrs of aerobic recovery (Fig. 9). However, early post-diapause II embryos (4 dpd, WS 36), which have the greatest ability to survive long-term anoxia, do not experience an increase in survival following anoxic preconditioning (Fig. 9). In fact, it appears that AP causes a statistically significant *decrease*

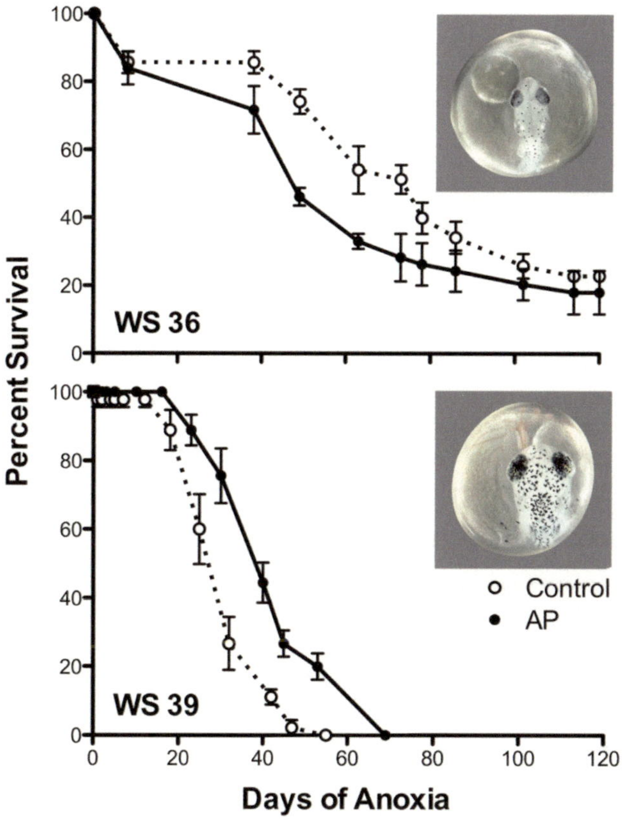

Fig. 9. Survival of embryos exposed to a single bout of long-term anoxia versus those exposed to anoxic preconditioning and then long-term anoxia at 25°C. Control = embryos incubated aerobically and then exposed directly to long-term anoxia; AP = embryos at same stage exposed to 24 hrs of anoxia and then 24 hrs of aerobic recovery prior to exposure to long-term anoxia. Embryos were exposed to anoxia as described in Podrabsky et al. (2007). Symbols represent the mean and error bars the S.E.M. (n=4 groups of 20 embryos).

in survival in this embryonic stage (Table 1). Later in development, as extreme tolerance of anoxia is lost, induction of endogenous protective mechanisms is observed (Fig 10). Both WS 39 and WS 40 embryos exhibit a significant increase in survival of anoxia following AP (Figs. 9 and 10). However, the extreme tolerance of anoxia observed in 4 dpd embryos cannot be induced in later stage embryos, suggesting that preconditioning is a unique anoxia tolerance phenotype. Escape embryos that do not enter diapause II appear to have an equal tolerance of anoxia compared to those that reach the same developmental stage after entering diapause II (Fig. 10). This suggests that AP-induced tolerance of anoxia is not a consequence of a diapause-induced trait that is partially retained or reactivated in a WS 40 embryo, but rather is a stage-specific trait.

		LT_{50}, days[a]	**95% C.I. of LT_{50}**			
Stage	**Preconditioning**	*Mean*	*Lower*	*Upper*	**% Change**	**Sig.**[b]
WS 36 - post-D2	None	74.3	65.9	82.7		A
	Anoxia	59.5	52.0	66.7	-20	B
WS 38 - post-D2	None	27.8	25.9	29.7		A
	Anoxia	39.1	37.1	41.1	41	B
WS 40 - post-D2	None	6.7	6.1	7.2		A
	Anoxia	8.8	8.3	9.4	32	B
	None	6.1	5.3	6.8		A
	Aerobic Acidosis	6.1	5.3	7.0	0	A
	None	5.4	4.8	6.0		A
	Hydrogen Sulfide	4.4	4.0	4.8	-18	C
WS 40- Escape	None	6.1	5.8	6.4		A
	Anoxia	9.4	8.9	10.0	56	B

[a]Probit regression analysis was used to determine the LT_{50} for each condition. LT_{50} values were compared for all experiments within a developmental stage, but not across developmental stages.
[b]LT_{50} values were considered statistically different if the 95% CI of their relative median potencies did not encompass 1. A relative median potency of 1 would indicate the same effect of each treatment on median survival. Comparisons were made for all treatments within a single developmental stage, but not between stages. Means with different letters are statistically different ($p<0.05$).

Table 1. 95% confidence intervals for lethal time to 50% mortality derived from Probit regression analysis of survivorship data.

The highly variable and extreme environment in which these embryos survive very likely imposes severe hypoxia or even anoxia on a daily basis (Podrabsky et al., 1998). Thus, the extreme tolerance observed in early embryos, and the ability to induce protective mechanisms in later embryos is very likely adaptive in this species. Given their habitat, it is curious that high anoxia tolerance would not be retained for all of development. This suggests that there may be some trade-off between maintaining high tolerance of anoxia and completion of normal development. It is possible that the protective mechanisms and metabolic alterations (see above) that support anoxia tolerance may partially interfere with rapid cell growth and proliferation, as has been reported for over-expression of stress proteins such as HSP 70 in *Drosophila melanogaster* (Krebs and Feder, 1997). There is certainly a consistent trend within *A. limnaeus* that periods of slow developmental progression and greatly reduced normoxic aerobic metabolism have a much greater tolerance of anoxia

(Podrabsky and Hand, 1999; Podrabsky et al., 2007). These slow periods of development are also associated with entry into diapause II and constitutively elevated levels of HSP 70 (Podrabsky and Somero, 2007). Thus, it is possible that these pathways may be at odds with fast rates of development, and thus maintenance of long-term anoxia tolerance comes at the cost of slower development.

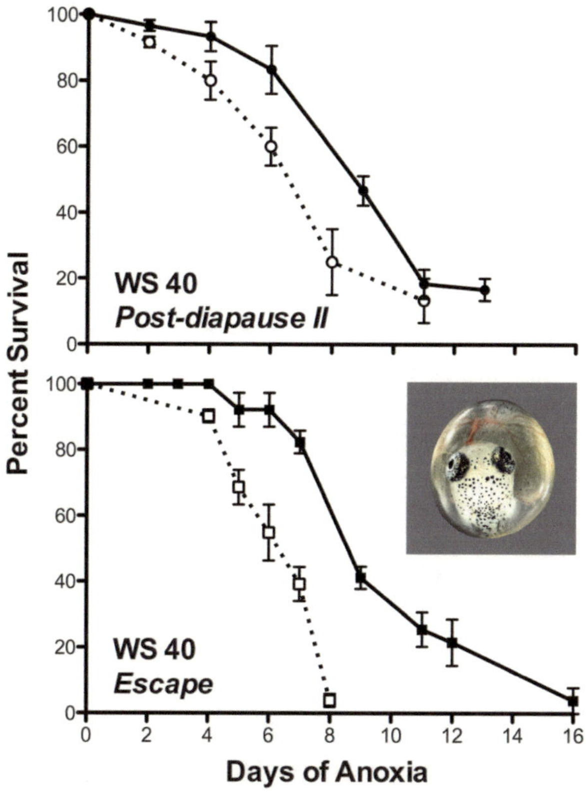

Fig. 10. Anoxic preconditioning (AP, closed symbols) can induce increased survival compared to control (open symbols) WS 40 embryos of *A. limnaeus* that have lost their long-term anoxia tolerance. Both embryos that entered diapause II and then resumed development, and those that escaped diapause II and reached the same stage have equivalent tolerances of anoxia and responses to AP. All experiments were conducted at 25°C. Symbols represent the mean ± S.E.M. (n=4 groups of 14-20 embryos).

7.2 Aerobic acidosis preconditioning

Aerobic acidosis is known to induce metabolic depression in other systems associated with long-term tolerance of anoxia. For instance, in brine shrimp, *Artemia franciscana*, aerobic acidosis can induce many of the phenotypic characters associated with metabolic depression (Hand and Carpenter, 1986). We see no affect of aerobic acidosis on increased survival following 20 hrs of preconditioning, although there was certainly a physiological affect as determined by decreased heart rate (Fig. 11). We conclude based on these two pieces of evidence that the mechanisms that reduce metabolism in aerobic acidosis are distinct from those that act during anoxia. This contrasts a number of studies in mammalian tissues that report a positive effect of acidotic preconditioning on survival of subsequent ischemic or hypoxic insults (Zhai et al., 1993; Lundmark et al., 1999; Luo et al., 2008). It is not clear why we do not see a similar effect in this species, but it could indicate that the mechanisms or signals that induce preconditioning in this species are novel and represent new avenues for extending cell survival during oxygen deprivation. It is also possible that because we did not allow the embryos to recover for 24 hrs following the aerobic acidosis, that they did not have enough time to induce the proper changes in gene expression. However, we find this unlikely because the embryos were always exposed to a normoxic environment, and would presumably still have plenty of resources available to support alterations in gene expression.

7.3 Sodium sulfide preconditioning

Recent evidence suggests that exposure to hydrogen sulfide can induce a state of metabolic depression in mice and can protect them from hypoxic damage (Blackstone and Roth, 2007). In addition, hydrogen sulfide signaling has been implicated as an oxygen sensing mechanism in the vasculature (Whitfield et al., 2008). Preconditioning *A. limnaeus* embryos with 500 μM sodium sulfide under aerobic conditions had no effect on survival of anoxia (Fig. 11), although it did have a clear physiological affect as evidenced by reduced heart rate. This suggests that sulfide signaling is likely not an important mediator of anoxic preconditioning in this system, but could be important in signaling metabolic depression through alternate metabolic pathways. This again points to a novel signaling mechanisms in *A. limnaeus* embryos compared to the typical mammalian models. However, this was a very simplified experimental regime and perhaps using a different level of sodium sulfide or allowing the embryos to recover aerobically for 24 hrs prior to exposing them to anoxia would alter the results.

7.4 Anoxic preconditioning in developing vertebrates

The concept of physiological preconditioning has been well-developed in response to short bouts of restricted blood flow (ischemia) in adult mammalian tissues and organs (e.g. Murray et al., 1986; Obrenovitch, 2008; Alchera et al., 2010; Chambers et al., 2010). This phenomenon is typified by decreased damage in oxygen sensitive tissues, increased physiological function, and increased survival of individuals suffering from an otherwise lethal dose of ischemia without prior preconditioning. These studies illustrate that short bouts of ischemia can induce a protective phenotype in a variety of tissues and organs of many mammalian species. However, most of these studies are focused on adult tissues and organs, and it is not clear that preconditioning is affective in early life history stages. For example, five 1 min ischemic events induced by umbilical occlusion did not result in a preconditioned phenotype in fetal sheep (Lotgering et al., 2004). Ostadalova et al. (1998) report no protective effects of ischemic preconditioning on isolated neonatal rat hearts until 7 days post-birth. They conclude that induction of protective preconditioning may not be developed until the baseline high tolerance typical of fetal and neonatal stages is lost (Ostadalova et al., 1998). Thus, the limited evidence suggests that while fetal and neonatal

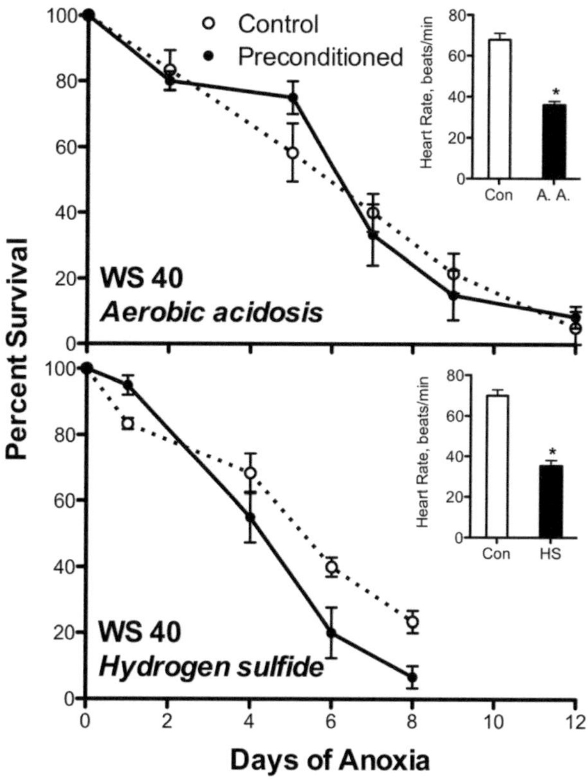

Fig. 11. Survival of *A. limnaeus* embryos at WS 40 that were exposed to long-term anoxia (Control) or were preconditioned with aerobic acidosis (20 hrs of exposure to 20% CO_2, 80% room air) or 500 μM sodium sulfide for 20 hrs under aerobic conditions. All experiments were conducted at 25°C. Symbols represent the mean and error bars the S.E.M. (n=4 groups of 20 embryos). The inset graphs for each panel report heart rate activity in beats per minute for each experimental treatment. Bars are means and error bars are S.D. (n=8 embryos). Both aerobic acidosis and sodium sulfide caused a significant decline in heart rate after 20 hrs of exposure (t-tests, $p<0.001$).

mammalian embryos have a higher overall tolerance of oxygen deprivation, they lack a strong preconditioning effect in heart tissue. However, asphyxic preconditioning does appear to induce increased survival in late term rat embryos (Strackx et al., 2010). This study exposed late gestation fetuses (embryonic day 17) to 30 min of ischemia by completely clamping the uterine and ovarian arteries. This treatment lead to a significant increase in survival in pups challenged with 19 min of severe asphyxia on embryonic day 22. The conflicting results reported in these two studies on rats may be due to differences in experimental methodologies. It is also possible that fetal and neonatal brain and heart tissues respond to preconditioning in different ways, and with a different time course of development. However, it is important to note that in the data presented in this study as well as in the few studies on mammalian fetal and neonatal stages, that preconditioning does not appear to re-establish the fetal/neonatal phenotype of substantially higher tolerance of anoxia, and thus the mechanisms that support tolerance of oxygen deprivation in early life history stages of vertebrates may be very different from the mechanisms that support protective preconditioning. This suggests that studies of both mechanisms could lead to novel insights into the prevention of damage due to oxygen deprivation.

7.5 Evolutionary significance of anoxic preconditioning

Ischemic or anoxic preconditioning has not been explored in many non-mammalian systems. Hypoxic preconditioning has been observed in the heart of hypoxia sensitive strains of trout, but not those that have naturally evolved an increased tolerance of hypoxia (Gamperl et al., 2001; Gamperl et al., 2004; Overgaard et al., 2004). These authors suggest that in the trout heart, hypoxia tolerance and preconditioning are not additive, which leads to the conclusion that hypoxia tolerance in this species may amount to the permanent induction of the preconditioned phenotype. This is not the case in embryos of *A. limnaeus*, which have a substantial tolerance of anoxia, even in late developmental stages and yet their survival can be extended through anoxic preconditioning. However, the present study and the work of Gamperl and colleagues (2001; 2004) suggest that anoxic or hypoxic preconditioning is very likely an ancient trait that evolved at least early in the vertebrate lineage. Indeed, it may be much more ancient than that. Johnson et al. (1989) report hypoxic preconditioning of anoxia tolerance in the root tips of corn. Perhaps the anoxic/hypoxic preconditioning response is simply the induction of very ancient stress tolerance pathways that for some reason are detrimental to multicellular organisms except under conditions that limit oxygen supply. The evolutionary history of the preconditioning phenomenon has not been comprehensively addressed, and detailed studies from a variety of lineages will be needed to assess the overall importance of this response to organismal survival of anoxia or hypoxia.

8. Conclusions

8.1 Anoxia tolerance during vertebrate development

Tolerance of anoxia is generally high during early development and is subsequently lost as vertebrate embryos develop towards the adult phenotype of low tolerance. This pattern applies to species from a diversity of lineages, reproductive strategies, and physiologies. The underlying reason for this pattern is currently not known, and cannot necessarily be explained by chronic hypoxia during development. Embryos of the annual killifish *Austrofundulus limnaeus* have an unequaled tolerance of anoxia among the vertebrates that is orders of magnitude higher even after developmental stage is taken into account. Despite this high tolerance, they respond to anoxia using the same basic metabolic pathways as more typical vertebrates that lack significant anoxia tolerance. The key to long-term survival of anoxia is the ability to enter a state of profound metabolic depression. Contrary to what is known in other anoxia tolerant vertebrates, ATP levels are not defended in anoxic embryos of *A. limnaeus*, which indicates a lack of tight coordination between energy production and consumption during transition into anoxia. This evidence suggests that embryos of *A. limnaeus* may have novel mechanisms of anoxia tolerance that have not yet been explored. A better understanding of these mechanisms has the potential to lead to new treatments for diseases and conditions caused by oxygen deprivation in humans.

8.2 Preconditioning during vertebrate development

The induction of preconditioning in developing vertebrates has received very little attention. The few studies that have been reported yield conflicting results depending on the species involved and likely the developmental stage and tissues investigated. In *A. limnaeus*, embryos with peak tolerance are already prepared for maximal survival of anoxia, and show no positive effects of anoxic preconditioning. Later in development, once this maximal tolerance of anoxia is lost, anoxic preconditioning does induce a protective phenotype. However, preconditioning does not result in recapitulating the lost long-term maximal tolerance of anoxia for this species, or in the other species investigated. This leads to the conclusion that there is something fundamentally different about the molecular phenotype of an embryo following anoxic preconditioning compared to those that naturally express an extremely high tolerance of anoxia as a consequence of their natural developmental program. This is also the case in mammalian systems, where preconditioning of adult tissues is unable to recapitulate the survival times of early life history stages. Thus, the mechanisms that support increased anoxia tolerance during vertebrate development and those that increase tolerance of anoxia due to preconditioning are likely unique. Continued investigation of both strategies has potential to yield a better understanding of anoxia tolerance in developing vertebrates and lead to new treatments for improving survival of oxygen deprivation in humans.

9. Acknowledgments

This research was supported by a NIH National Heart Lung and Blood Institute grant to J.E.P (R01HL095454).

10. References

Adolph, E.F. (1969). Regulations during survival without oxygen in infant mammals. *Respiration Physiology*, 7, 356-368.

Adolph, E.F. (1979). Development of dependence on oxygen in embryo salamanders. *American Journal of Physiology*, 236, R282-R291.

Alchera, E., Dal Ponte, C., Imarisio, C., Albano, E., & Carini, R. (2010). Molecular mechanisms of liver preconditioning. *World Journal of Gastroenterology*, 16, 6058-6067.

Ben-Ari, Y. (2002). Excitatory actions of gaba during development: the nature of the nurture. *Nature Reviews Neuroscience*, 3, 728-739.

Bharmaa S., & Milsom, W.K. (1993). Acidosis and metabolic rate in golden mantled ground squirrels (Spermophilus lateralis). *Respiration Physiology*, 94, 337-351.

Blackstone, E., & Roth, M.B. (2007). Suspended animation-like state protects mice from lethal hypoxia. *Shock*, 27, 370-372.

Cai, Z., Fratkin, J.D., Rhodes, P.G. (1997). Prenatal ischemia reduces neuronal injury caused by neonatal hypoxia-ischemia in rats. *NeuroReport*, 8, 1393-1398.

Ceylan, H., Yuncu, M., Gurel, A., Armutcu, F., Gergerlioglu, H.S., Bagci, C., & Demiryurek, A.T. (2005). Effects of whole-body hypoxia preconditioning on hypoxia/reoxygenation-induced intestinal injury in newborn rats. *European Journal of Pediatric Surgery*, 15, 325-332.

Chambers, D.J., Fallouh, H.B., & Kassem, N.A. (2010). Ischemic preconditioning and lung preservation, In: *Principles of Pulmonary Protection in Heart Surgery*, Gabriel, E.A., & Salerno, T., pp. (223-234), Spinger-Verlag, London.

Chennault, T., & Podrabsky, J.E. (2010). Aerobic and Anaerobic Capacities Differ in Embryos of the Annual Killifish *Austrofundulus limnaeus* that develop on Alternate Developmental Trajectories. *Journal of Experimental Zoology A*, 313A, 587-596.

Cherubini, E., Gaiarsa, J.L., & Ben-Ari, Y. (1991). GABA: an excitatory transmitter in early postnatal life. *Trends in Neurosciences,* 14, 515-519.

Cohen, M.V., Yang, X.M., & Downey, J.M. (1994). Conscious rabbits become tolerant to multiple episodes of ischemic preconditioning. *Circulation Research*, 74, 998-1004.

Daniel, J.C.Jr. (1968). Oxygen concentrations for culture of rabbit blastocysts. *Journal of Reproduction and Fertility,* 17, 187-190.

Dawes, G.S., Mott, J.C., & Shelley, H.J. (1959). The importance of cardiac glycogen for the maintenance of life in foetal lambs and new-born animals during anoxia. *Journal of Physiology,* 146, 516-538.

Dawes, G.S., Jacobson, H.N., Mott, J.C., & Shelley, H.J. (1960). Some observations on foetal and new-born Rhesus monkeys. *Journal of Physiology,* 152, 271-298.

Degos, V., Loron, G., mantz, J., & Gressens, P. (2008). Neuroprotective strategies for the neonatal brain. *Anesthesia & Analgesia,* 106, 1670-1680.

Di Carlo, A.L., & Litovitz, T.A. (1999). Is genetic the unrecognized confounding factor in bioelectromagnetics? Flock-dependence of field-induced anoxia protection in chick embryos. *Bioelectrochemistry and Bioenergetics*, 48, 209-215.

Dinkelacker, S.A., Costanzo, J.P., Lee, R.E.Jr. (2005). Anoxia tolerance and freeze tolerance in hatchling turtles. *Journal of Comparative Physiology B,* 175, 209-217.

DiMichele, L, & Powers, D.A. (1984). The relationship between oxygen consumption rate and hatching in *Fundulus heteroclitus. Physiological Zoology*, 57, 46-51.

Duerr, J.M., & Podrabsky, J.E. (2010). Mitochondrial physiology of diapausing and developing embryos of the annual killifish *Austrofundulus limnaeus*: Implications for extreme anoxia tolerance. *Journal of Comparative Physiology B*, 180, 991-1003.

Fazekas, J.F., Alexander, F.A.D., & Himwich, H.E. (1941). Tolerance of the newborn to anoxia. *American Journal of Physiology,* 134, 281-287.

Fergusson-Kolmes, L., & Podrabsky, J.E. (2007). Differential effects of anoxia on heart activity in anoxia-tolerant and anoxia-sensitive embryos of the annual killifish *Austrofundulus limnaeus. Journal of Experimental Zoology A,* 307A, 419-423.

Gamperl, A.K., Todgham, A.E., Parkhouse, W.S., Dill, R., & Farrell, A.P. (2001). Recovery of trout myocardial function following anoxia: preconditioning in a non-mammalian model. *American Journal of Physiology Regulatory Integrative and Comparative Physiology,* 281, R1755-R1763.

Gamperl, A. K., Faust, H.A., Dougher, B., & Rodnick,K.J. (2004). Hypoxia tolerance and preconditioning are not additive in trout (*Oncorhynchus mykiss*) heart. *Journal of Experimental Biology*, 207, 2497-2505.

Gidday, J.M. (2006). Cerebral preconditioning and ischaemic tolerance. *Nature Reviews Neuroscience,* 7, 437-448.

Glass, H.G., Snyder, F.F., & Webster, E. (1944). The rate of decline in resistance to anoxia of rabbits, dogs and guinea pigs from the onset of viability to adult life. *American Journal of Physiology,* 140, 609-615.

Hall, R. (1925). Effects of oxygen and carbon dioxide on the development of the whitefish. *Ecology,* 6, 104-116.

Hand, S.C. (1998). Quiescence in *Artemia franciscana* embryos: reversible arrest of metabolism and gene expression at low oxygen levels. *Journal of Experimental Biology,* 201, 1233-1242.

Hand, S.C., & Carpenter, J.F. (1986) pH-Induced Metabolic Transitions in *Artemia* Embryos Mediated by a Novel Hysteretic Trehalase. *Science,* 232, 1535-1537.

Hand, S.C., & Hardewig, I. (1996). Downregulation of cellular metabolism during environmental stress: mechanisms and implications. *Annual Review of Physiology,* 58, 539-563.

Hochachka, P.W., & Somero, G.N. (2002). *Biochemical Adaptation: Mechanism and Process in Physiological Evolution,* Oxford University Press, New York.

Ingalls, T.H., Curley, F.J., & Prindle, R.A. (1950). Anoxia as a cause of fetal death and congenital defect in the mouse. *American Journal of Diseases of Children,* 80, 34-45.

Johnson, J., Cobb, G., & Drew, M.C. (1989). Hypoxic induction of anoxia tolerance in root tips of *Zea mays*. *Plant Physiology,* 91, 837-841.

Johnston, I.A., & Bernard, L.M. (1983). Utilization of the ethanol pathway in carp following exposure to anoxia. *Journal of Experimental Biology,* 104, 73-78.

Kabat, H. (1940). The greater resistance of very young animals to arrest of the brain circulation. *American Journal of Physiology,* 130, 588-599.

Krebs, R.A., & Feder, M.E. (1997). Deleterious consequences of Hsp70 overexpression in Drosphilla melanogaster larvae. *Cell Stress Chaperones,* 2, 60–71.

Krumschnabel, G. (2000). Cellular and molecular basis of anoxia-tolerance and-intolerance in vertebrates: comparative studies using hepatocytes from goldfish and trout. *Recent Research Developments in Comparative Biochemistry and Physiology,* 1, 1-11.

Kültz, D. (2003). Evolution of the cellular stress proteome: from monophyletic origin to ubiquitous function. *Journal of Experimental Biology,* 206, 3119-3124.

Kültz, D. (2005). Molecular and evolutionary basis of the cellular stress response. *Annual Review of Physiology,* 67, 225-257.

Loeb, J. (1894). Ueber die relative empfindlichkeit von fischembryonen genen sauerstoffmangel und wasserentziehung in verschiedenen enteicklungsstadien. *Pflügers Archiv European Journal of Physiology,* 55, 530-541.

Longo, L.D., & Packianathan, S. (1997). Hypoxia-ischaemia and the developing brain: hypotheses regarding the pathophysiology of fetal-neonatal brain damage. *BJOG: An International Journal of Obstetrics & Gynaecology,* 104, 652-662.

Lotgering, F.K., Bishai, J.M., Struijk, P.C., Blood, A.B., Hunter, C.J., Oberg, K.C., Power, G.G., & Longo, L.D. (2004). Absence of robust ischemic preconditioning by five 1-minute total umbilical cord occlusions in fetal sheep. *Journal of the Society for Gynecologic Investigation,* 11, 449-456.

Lundmark, J.A., Trueblood, N., Wang L.F., Ramasamyf, R., & Schaefer, S. (1999). Repetitive acidosis protects the ischemic heart: implications for mechanisms in preconditioned hearts. *Journal of Molecular and Cellular Cardiology*, 31, 907-917.

Luo, H., Chang, Y., Cai, H., Zou, W., Wng, D., & Guo, Q. (2008). The effect of hypercapnic acidosis preconditioning on rabbit myocardium. *Journal of Huazhong University of Science and Technology -- Medical Sciences*, 28, 706-710.

Mendelsohn, B.A., Kassebaum, B.L., & Gitlin, J.D. (2008). The zebrafish embryo as a dynamic model of anoxia tolerance. *Developmental Dynamics*, 237, 1780-1788.

Mueller, C.A., Joss, J.M.P., & Seymour, R.S. (2011). Effects of environmental oxygen on development and respiration of Australian lungfish (*Neoceratodus forsteri*) embryos. *Journal of Comparative Physiology B*, 181, 941-952.

Murray C.E., Jennings, R.B., & Reimer, K.A. (1986). Preconditioning with ischemia: a delay of lethal cell injury in ischemic myocardium. *Circulation*, 74, 1124-1136.

Mulvey, J.M., & Renshaw, G.M.C. (2000). Neuronal oxidative hypometabolism in the brainstem of the epaulette shark (*Hemiscyllium ocellatum*) in response to hypoxic pre-conditioning. *Neuroscience Letters*, 290, 1-4.

Nelson, O.E. (1958). The development of the primitive streak and early chick embryos in relation to low O_2 pressure. *Growth*, 22, 109-124.

Nilsson, G.E., & Lutz, P.L. (2004). Anoxia tolerant brains. *Journal of Cerebral Blood Flow & Metabolism*, 24, 475-486.

Obrenovitch, T.P. (2008). Molecular physiology of preconditioning-induced brain tolerance to ischemia. *Physiological Reviews*, 88, 211-247.

Ostadal, B., Ostadalova, I., & Dhalla, N.S. (1999). Development of cardiac sensitivity to oxygen deficiency: comparative and ontogenetic aspects. *Physiological Reviews*, 79, 635-659.

Ostadalova, I., Ostadal, B., Kolar, F., Parratt, J.R., & Wilson, S. (1998). Tolerance to ischaemia and ischaemic preconditioning in neonatal rat heart. *Journal of Molecular and Cellular Cardiology*, 30, 857-865.

Overgaard, J., Stecyk, J.A.W., Gesser, H., Wang, T., Gamperl, A.K., & Farrell, A.P. (2004). Preconditioning stimuli do not benefit the myocardium of hypoxia-tolerant rainbow trout (*Oncorhynchus mykiss*). *Journal of Comparative Physiology B*, 174, 329-340.

Podrabsky, J.E. (1999). Husbandry of the annual killifish *Austrofundulus limnaeus* with special emphasis on the collection and rearing of embryos. *Environmental Biology of Fishes*, 54, 421-431.

Podrabsky, J.E., & Hand, S.C. (1999). The bioenergetics of embryonic diapause in an annual killifish, *Austrofundulus limnaeus*. *Journal of Experimental Biology*, 202, 2567-2580.

Podrabsky, J.E., & Somero, G.N. (2007). An inducible 70 kDa-class heat shock protein is constitutively expressed during early development and diapause in the annual killifish *Austrofundulus limnaeus*. *Cell Stress and Chaperones*, 12, 199-204.

Podrabsky, J.E., Hrbek, T., & Hand, S.C. (1998). Physical and chemical characteristics of ephemeral pond habitats in the Maracaibo basin and Llanos region of Venezuela. *Hydrobiologia*, 362, 67-78.

Podrabsky, J.E., Lopez, J.P., Fan, T.W.M., Higashi. R., & Somero, G.N. (2007). Extreme anoxia tolerance in embryos of the annual killifish *Austrofundulus limnaeus*: Insights from a metabolomics analysis. *Journal of Experimental Biology,* 210, 2253-2266.

Podrabsky, J.E., Garrett, I.D.F., & Kohl, Z.F. (2010a). Alternate developmental pathways associated with diapause in embryos of the annual killifish *Austrofundulus limnaeus. Journal of Experimental Biology,* 213, 3280-3288.

Podrabsky, J.E., Tingaud-Sequeira, A., & Cerda, J. (2010b). Metabolic Dormancy and Responses to Environmental Desiccation in Fish Embryos. *Topics in Current Genetics,* 21, 203-226.

Riddle, O. (1924). On the necessary gaseous environment of the bird embryo. *Ecology,* 5, 348-362.

Rink, R., & Miller, J.A.Jr. (1967). Temperature, weight (=age) and resistance to asphyxia in pouch-young opossums. *Cryobiology,* 4, 24-29.

Rose, F.L., Armentrout, D., & Roper, P. (1971). Physiological responses of paedogenic *Ambystoma tigrinum* to acute anoxia. *Herpetologica,* 27, 101-107.

Sedmera, D., Kucera, P., & Raddatz, E. (2002). Developmental changes in cardiac recovery from anoxia-reoxygenation. *American Journal of Physiology Regulatory Integrative and Comparative Physiology,* 283, R379-R388.

Shepard, M.P. (1955). Resistance and tolerance of young speckled trout (*Salvelinus fontinalis*) to oxygen lack, with special reference to low oxygen acclimation. *Journal of the Fisheries Research Board of Canada,* 12, 387-446.

Shoubridge, E.A., & Hochachka, P.W. (1980). Ethanol: novel end product of vertebrate anaerobic metabolism. *Science,* 209, 308-309.

Strackx, E., Van den Hove, D.L.A., Prickaerts, J., Zimmermann, L., Steinbusch, H.W.M., Blanco, C.E., Gavilanes, A.W.D., & Vles, J.S. Hans. (2010). Fetal asphyctic preconditioning protects against perinatal asphyxia-induced behavioral consequences in adulthood. *Behavioural Brain Research,* 208, 343-351.

Tyzio, R., Cossaart, R., Khalilov, I., Minlebaev, M., Hübner, C.A., Represa, A., Ben-Ari, Y., & Khazipov, R. (2006). Maternal oxytocin triggers a transient inhibitory switch in GABA signaling in the fetal brain during delivery. *Science,* 314, 1788-1792.

van Waarde, A. (1991). Alcoholic fermentation in multicellular organisms. *Physiological Zoology,* 64, 895-920.

Verdouw, P.D., van den Doel, M.A., de Zeeuw, S., & Duncker, D.J. (1998). Animal models in the study of myocardial ischaemia and ischaemic syndromes. *Cardiovascular Research,* 39, 121-135.

Weigmann, A.L., & Altig, R. (1975). Anoxic tolerances of three species of salamander larvae. *Comparative Biochemistry and Physiology A,* 50A, 681-684.

Whitfield, N.L., Kreimier, E.L., Verdial, F.C., Skovgaard, N., & Olson, K.R. (2008). Reappraisal of the H_2S/sulfide concentration in vertebrate blood and its potential significance in ischemic preconditioning and vascular signaling. *American Journal of Physiology Regulatory Integrative and Comparative Physiology,* 294, R1930-R1937.

Wourms, J.P. (1972a). Developmental biology of annual fishes I. Stages in the normal development of *Austrofundulus myersi* Dahl. *Journal of Experimental Zoology,* 182, 143–167.

Wourms, J.P. (1972b) The developmental biology of annual fishes III. Pre-embryonic and embryonic diapause of variable duration in the eggs of annual fishes. *Journal of Experimental Zoology,* 182, 389-414.

Zhai X, Lawson C.S., Cave A.C., & Hearse, D.J. (1993). Preconditioning and post-ischaemic contractile dysfunction: the role of impaired oxygen delivery vs extracellular metabolite accumulation. *Journal of Molecular and Cellular Cardiology,* 25, 847-857.

2

Anoxia-Induced Suspended Animation in *Caenorhabditis elegans*

Pamela A. Padilla, Jo M. Goy and Vinita A. Hajeri
University of North Texas, Department of Biological Sciences
Denton, TX
USA

1. Introduction

Some of the most complex biological processes were first elucidated in somewhat "simple" genetic model organisms. For example, where would we be in our molecular and cellular understanding of gene expression, cell division, embryo development and cell death if it were not for research using *E. coli*, Yeast, *Drosophila* and *C. elegans*? Due to the pioneering work by Sydney Brenner and others, the soil nematode *C. elegans* is now a well-known genetic and developmental model system (Brenner, 1974). Genetic approaches have contributed significantly to our advanced understanding of the mechanisms regulating gene function, organ development, microRNA function and signaling pathways regulating aging and stress (WormBook). The molecular advances and development of genetic tools such as RNA interference (RNAi) and protein expression analysis with the Green Fluorescent Protein (GFP) were initially worked out in *C. elegans* and thus firmly established this organism as a cornerstone of genetic models for unraveling the molecular mechanisms of many biological processes (Chalfie and Kain, 2006; Fraser et al., 2000; Jorgensen and Mango, 2002; Timmons and Fire, 1998). Research from many labs have clearly demonstrated that this small soil nematode has contributed significantly to our understanding of biology and that multiple types of molecular tools exist to elucidate mechanistic details. Further examples of the significant impact *C. elegans* has had in our understanding of biology are the fairly recent Noble Prize awards to six individuals (S. Brenner, M. Chalfie, A. Fire, R. Horvitz, C. Mello, J. Sulston) who made paradigm shifting discoveries using *C. elegans*. Through the effort of many within the *C. elegans* community, the molecular tools and genetic resources available in this model system have helped to address and elucidate the molecular mechanisms regulating multiple biological processes.

C. elegans, a soil nematode, was originally isolated in Bristol, England (N2 Bristol strain) (Brenner, 1974). In its natural environment oxygen levels fluctuate, therefore *C. elegans* likely evolved mechanisms to survive changes in oxygen levels (Lee, 1965). Indeed, it is known that *C. elegans* survive oxygen deprivation (hypoxia and anoxia) that have an impact on behavior, growth and development (Anderson, 1978; Padilla et al., 2002; Paul et al., 2000; Van Voorhies and Ward, 2000). *C. elegans* are able to sense various oxygen tensions and in fact prefer a 5-12% O_2 concentration (Gray et al., 2004). A lower level of oxygen in the environment is stressful to the worm yet the animal has evolved mechanisms to survive the stress of hypoxia and anoxia. A review by Powell-Coffman nicely provides an overview of the signaling pathways and responses involved with oxygen deprivation in *C. elegans* (Powell-Coffman, 2010). In *C. elegans* and other metazoans a key to sensing low oxygen levels within the environment is the hypoxia-inducible factor HIF-1. The transcription factor HIF-1 is needed to induce the expression of a variety of genes so that the animal may survive low oxygen environments (Jiang et al., 2001). The mechanisms regulating HIF-1 activity are being worked out in a variety of systems including *C. elegans* and has been reviewed in many publications (Epstein et al., 2001; Semenza, 2007; Semenza, 2010) Interestingly, for *C. elegans*, the level of oxygen in the environment will dictate which genes are required for oxygen deprivation response and survival. For example, HIF-1 function is required for animals to survive and maintain normal developmental functions in .5% and 1% O_2 however, HIF-1 is not required for anoxia survival (Padilla et al., 2002). This review will focus on the response to anoxia and known mechanisms required for *C. elegans* to survive anoxia.

2. Methodology for studying anoxia response and survival in *C. elegans*

Given that the specific oxygen concentration affects the response *C. elegans* has to the environment it is important to discuss the methodologies used to produce an anoxic environment within the laboratory setting. Typically, several laboratory methods are used to expose *C. elegans* to anoxia. Figure 1 provides a general review of the methodologies commonly used and the pros and cons of each method. For example, a convenient and cost effective approach is to place *C. elegans*, which are grown on agar nematode growth media (NGM), into anaerobic biobags (Becton Dickson Company). These anaerobic biobags are typically used for growing anaerobic bacteria and use resazurin at an oxygen indicator. Another approach that allows one to study subcellular responses to brief periods of anoxia, in live animals, is to use an anoxia flow-through chamber in conjunction with a spinning disc confocal microscope. This method has been valuable in following chromosome structure in embryos and oocytes within adult hermaphrodites exposed to brief periods of anoxia. If one is planning to expose a large number of animals to anoxia, the use of a hypoxia chamber that can hold many *C. elegans* plates is the best approach. The use of a hypoxia chamber has been valuable for large-scale forward genetic or RNA interference screens. There are different types of hypoxia chambers; those that can be commercially purchased or those that can be tailor designed by the researcher. The chambers that are researcher designed can be made to allow a flow through of nitrogen to replace the air. The commercially available glove box chambers (Ex: Ruskinn Inc. anaerobic and hypoxia workstations) are useful if one needs to expose animals to oxygen deprivation for very long periods of time or if the animals need to be manipulated while exposed to anoxia. The primary disadvantage to the glove box chambers is that for some models a temperature below ambient can only be reached if the entire chamber is located within a low temperature room. Temperature during anoxia exposure is an important consideration given that temperature does influence anoxia survival rate (LaRue and Padilla, 2011). It is of interest to use more than one of these methodologies to verify observed anoxia responses and phenotypes are consistent.

Method	Pros	Cons
Anaerobic biobag 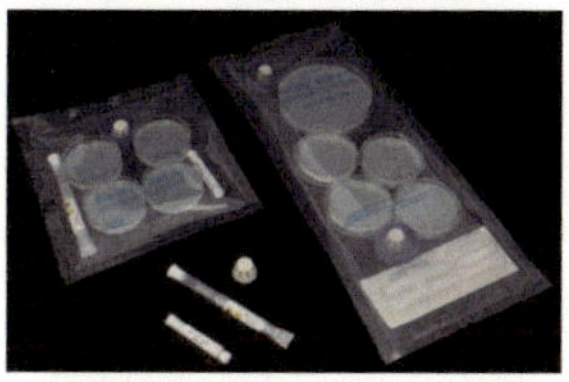	Inexpensive, commercially available Portable and can be placed in various temperature incubators Fairly consistent transition time to anoxia No need for gas tanks	Limited number of plates can be put into the environment A brief increase in temperature due to the chemical reaction used to remove oxygen within the biobag
Microscope chamber 	Visualize subcellular and cellular changes *In vivo* analysis of GFP fusion protein markers Hypoxia experiments also feasible	Exposure time is limited Potential issues such as sample dehydration Small number of animals analyzed
Flow-through chamber 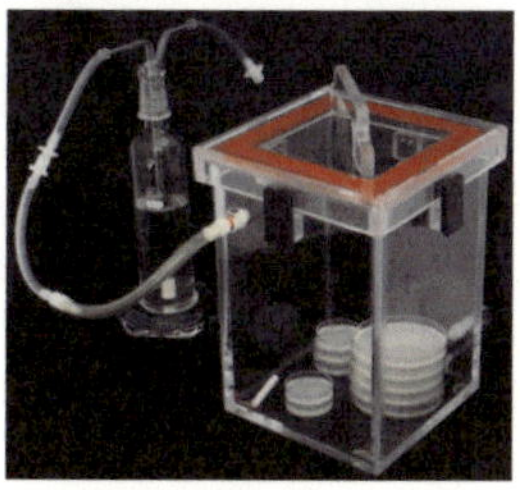	Researcher designed for specific experiments Hypoxia experiments also feasible Large number of animals can be simultaneously exposed Can be placed in a specific temperature controlled room	Transition time from normoxia to anoxia can vary
Glove box chamber 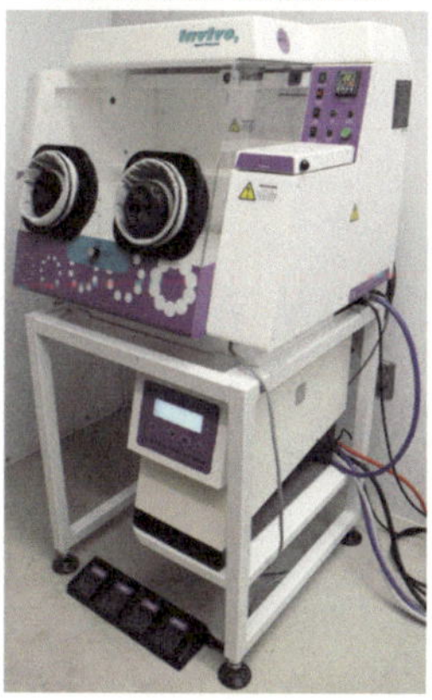	Hypoxia experiments also feasible Manipulation of animals while in the anoxic environment Can immediately place animals into the environment with no transition time Large number of animals can be simultaneously exposed Commercially available from various sources	Costly Chamber temperature higher than ambient unless placed in room with reduced temperature

Table 1. Various methodologies used to expose *C. elegans* to anoxia.

3. *C. elegans* as a model to study anoxia-induced suspended animation

Some organisms, including metazoans, exposed to stresses or naturally occurring signals can arrest processes such as development, cell division or heartbeat (Clegg, 2001; Mendelsohn et al., 2008; Padilla and Roth, 2001; Podrabsky et al., 2007; Renfree and Shaw, 2000; Riddle, 1988). Suspended animation is the arrest of observable biological processes induced by either a cue or stress in the environment, or a signal from within the animal. In the case of *C. elegans*, animals exposed to anoxia will enter into a reversible state of suspended animation in which observable biological processes, such as cell division, development, eating, egg laying, fertilization and movement arrests until air is reintroduced into the environment. Suspended animation can be maintained for a few days, depending on the developmental stage of exposure; extended periods of anoxia exposure will lead to lethality. Figure 1 demonstrates the developmental arrest that is observed in *C. elegans* exposed to anoxia. In a hypometabolic state, such as suspended animation, homeostasis is maintained until the environment required to support energy requiring processes is resumed. However, hypometabolic states, including suspended animation, are not maintained indefinitely and at some point the animal may die if the environment is not shifted to a more conducive state to support biological processes. Embryonic diapause, another hypometabolic state that can be either obligate or environmentally induced, is a natural survival strategy to maintain populations and maximize offspring. Embryonic diapause can be thought of as a state of suspended animation in that development and cell divisions are arrested (Clegg, 2001). An example of a vertebrate that enters into an obligate diapause is the killifish embryo *Austrofundulus limnaeus*. *A. limnaeus* embryos in diapause II are remarkably tolerant to anoxia (Podrabsky et al., 2007). The mechanistic overlap between developmental arrests induced by anoxia in comparison to naturally occurring diapause remains to be determined.

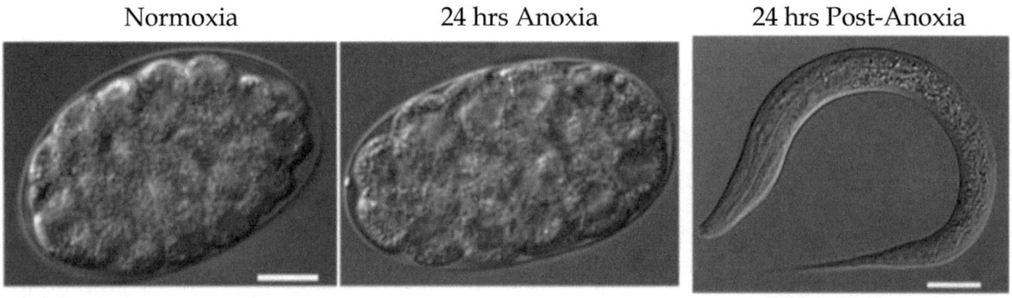

Fig. 1. *C. elegans* exposed to anoxia enter into a reversible state of suspended animation in which development and cell cycle progression will arrest. Shown are embryos collected from a gravid adult and exposed to normoxia or anoxia for 24 hours. The anoxia-exposed embryo will arrest development for several days and yet remain viable (shown is an embryo exposed to 24 hours of anoxia). The post-anoxia animal will resume development when air is added back to the environment. Scale bar = 10 μm for embryos and 20 μm for post-anoxia larva.

At every stage of development *C. elegans* survive 24 hours of anoxia exposure at a rate of 90% or greater (Foll et al., 1999; Padilla et al., 2002; Van Voorhies and Ward, 2000). The most anoxia tolerant developmental stages are dauer larvae and embryos (Anderson, 1978;

Padilla et al., 2002). The ability to survive longer bouts of anoxia depends upon developmental stage, growth temperature, diet, genotype and fertility; these factors will be elaborated on further in this chapter. In general, *C. elegans* are sensitive to anoxia (and hypoxia) if the temperature during exposure is increased (Ex: 28°C instead of 20°C) or if the duration of anoxia exposure is increased from one to three days (Mendenhall et al., 2006; Scott et al., 2002). After non-lethal exposures to anoxia the animals will resume biological processes such as cell division, development, eating, movement, and offspring production. How quickly animation resumes is dependent upon the anoxia exposure time. For example, an embryo that was exposed to one day of anoxia will resume cell cycle progression faster than an embryo that was exposed to three days of anoxia (Hajeri et al., 2005). Many aspects of anoxia response and survival are not understood. Listed below are questions that will be of interest to address in terms of molecular mechanisms regulating anoxia responses.

1. What genetic factors control entry, maintenance and exit from anoxia-induced suspended animation?
2. What cellular changes occur in animals exposed to anoxia and are such changes necessary and sufficient for anoxia-induced suspended animation?
3. In the embryo, how does a reduction in oxygen levels signal cell cycle arrest; via which cell cycle machinery?
4. How are developmental programs arrested and resumed in embryos and larvae exposed to anoxia?
5. How are complex tissues, such as muscles and neurons in adults, maintained during anoxia exposure?
6. What is the metabolic state of animals exposed to anoxia relative to duration and developmental state?
7. How do metabolic changes influence anoxia-induced suspended animation?
8. What molecular mechanisms balance offspring production and anoxia stress survival?
9. How can anoxia studies in *C. elegans* be used to better understand oxygen deprivation sensitivity in humans?

4. Anoxia-induced cell cycle arrest in the developing embryo

There are known environmental changes that influence cell cycle progression. For example, UV radiation will activate cell cycle checkpoint proteins and lead to a cell cycle arrest and repair of DNA damage so that the cell can progress through cell division (Hartwell and Weinert, 1989; Nurse et al., 1998). Also, exposing cells to drugs (Ex: Taxol, Nocodazole) was instrumental in the identification of cell cycle checkpoint genes. Identifying the fundamental regulation of cell cycle progression is central to the development of cancer treatments, thus understanding how oxygen levels affect cell division is of interest. Anoxia-exposed *C. elegans* embryos contain blastomeres that arrest at specific positions of the cell cycle: interphase, late prophase and metaphase. The lack of anaphase blastomeres indicates that the embryos are not progressing through the cell cycle further supporting that these embryos are indeed arrested. The phenomenon of anoxia-induced arrest is not unique to *C. elegans* since zebrafish (*Danio rerio*) and *Drosophila melanogaster* embryos also arrest cell cycle progression in response to anoxia or hypoxia (DiGregorio et al., 2001; Douglas et al., 2001; Foe and Alberts, 1985; Padilla and Roth, 2001). The use of cell biological techniques, such as indirect immunofluroescent assays or *in vivo* GFP fusion protein assays, showed that anoxia-arrested blastomeres have specific characteristics or hallmarks (Hajeri et al., 2005). An overview of the anoxia-induced cellular changes observed in embryos is summarized in Table 2 and discussed in detail throughout this section.

Cell cycle stage and environment	Characteristics of anoxia-arrested blastomeres and genes required
Interphase Nucleus: Normoxia Anoxia 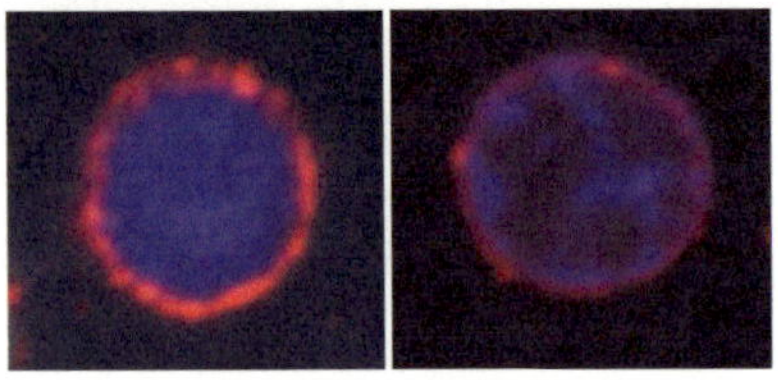	• Chromatin is condensed instead of dispersed throughout nucleus • Chromatin associates with inner nuclear membrane • Based on centrosome location, a specific stage of interphase arrest is not observed • Genes required for interphase arrest: none identified to date
Late Prophase Nucleus: Normoxia Anoxia 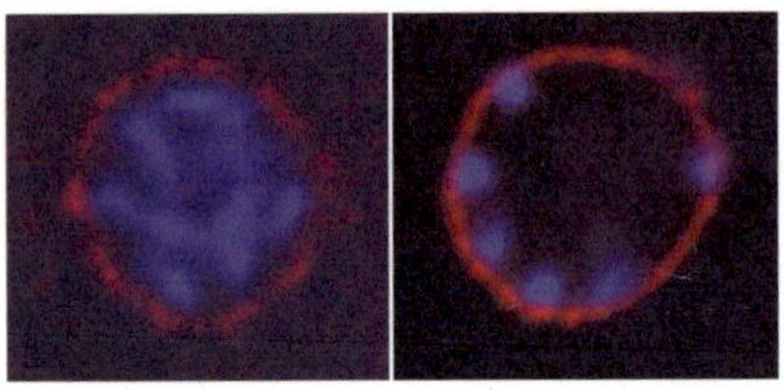	• Chromosomes fully condensed • Chromosomes "dock" near the inner nuclear membrane • CDK-1 localizes to the chromosomes • Inactive form of CDK-1 (P^{Tyr15}) is detected • Genes required for prophase arrest: *npp-16* (NUP-50 homologue which is a component of the nuclear pore complex) and *hda-2* (histone deacetylase)
Metaphase Nucleus: Normoxia Anoxia 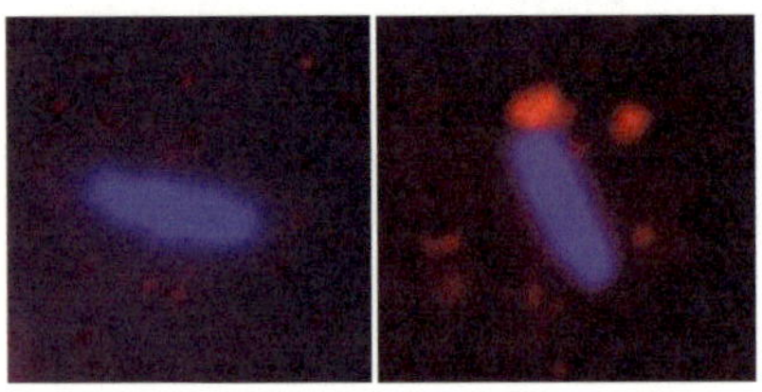	• Nuclear pore complex aggregates, as detected by mAb 414 • Spindle microtubules are present but show some depolymerization • Spindle checkpoint protein SAN-1 localization altered • Reduced phosphorylation of various proteins including Histone H3 at Serine 10 and proteins recognized by MPM-2 antibody • Genes required: Spindle checkpoint genes *san-1* and *mdf-2*

Table 2. Characteristics of anoxia-induced cell cycle arrest. Shown are representative nuclei from normoxic and anoxic embryos. Nuclear membrane, recognized by mAb 414, is shown in red and DNA, recognized by DAPI, is shown in blue.

4.1 Cellular changes associated with interphase arrest

Interphase blastomeres of anoxia-exposed embryos contain chromatin that is highly condensed and the level of condensation appears to increase with longer periods of anoxia exposure (Foe and Alberts, 1985; Hajeri et al., 2005). Chromatin condensation is characteristic of inactive chromatin and thus it is likely that a global down regulation of gene expression occurs in anoxic embryos. A reduction in gene expression is likely a means to conserve energy and maintain metabolic homeostasis (Hochachka et al., 1996). We know little about the mechanisms that regulate arrest of interphase blastomeres and if the interphase blastomeres arrest at a specific position of interphase (G1, S or G2). A challenge with *C. elegans* embryos is that the onset of gap phases may be lineage dependent. Thus, cell lineage would likely need to be considered when trying to determine the position of interphase arrest.

4.2 Cellular changes and genes associated with metaphase arrest

Metaphase arrest or delay has been studied in other systems such as yeast or vertebrate cells in culture exposed to microtubule inhibitors. Through these studies the spindle checkpoint pathway and a greater understanding of cell cycle progression has been elucidated. The advantage of investigating anoxia-induced metaphase arrest in *C. elegans* embryos is that this can be studied in a developing organism and that oxygen deprivation is a stress that the organism must have adapted to in nature. The cellular changes observed in anoxia-arrested metaphase blastomeres are influenced by anoxia exposure time. For example, depolymerization of astral and spindle microtubules and a reduction of the spindle checkpoint protein SAN-1 at the kinetochore is more extensive in embryos exposed to three days of anoxia in comparison to one day (Hajeri et al., 2005). Likewise, there is a reduction in phosphorylation of certain proteins such as Histone H3 at Serine 10 and the mitotic proteins recognized by mAb MPM-2 (Padilla et al., 2002). These cellular changes could be due to a decrease in energy levels, in the form of ATP, with increased anoxia exposure time.

In many organisms including *C. elegans*, the mAb 414 recognizes FG repeats of specific nucleoporin proteins that are components of the nuclear pore complex (NPC) (Lee et al., 2000). In normoxic embryos, mAb 414 will recognize the NPC of interphase, prophase and prometaphase blastomeres; mAb 414 signal is diminished in metaphase and anaphase blastomeres and will reform in telophase blastomeres. Therefore, mAb 414 is an excellent marker for the NPC and cell cycle position. In anoxic embryos, the metaphase blastomere contains NPC aggregates recognized by mAb 414. The significance of these NPC aggregates is not known, but is a consistent characteristic of anoxia-arrested metaphase blastomeres. All of these anoxia-induced cellular changes are reversible when the embryos are re-exposed to normoxia. The arrested metaphase blastomere will transition to anaphase and chromosome segregation will take place.

An RNAi screen for genes on chromosome I that when knocked-down lead to an anoxia sensitivity phenotype showed that the spindle checkpoint is required for anoxia-induced metaphase arrest (Nystul et al., 2003). RNAi or genetic knockdown of the spindle checkpoint genes, *san-1* (mad-3/BubR1 homologue) and *mdf-1* (mad-1 homologue) leads to a decrease in the viability of embryos exposed to anoxia. The *san-1(RNAi)* as well as the *san-1(ok1580)* deletion mutant are sensitive to anoxia exposure. These embryos contain a dramatic decrease in the number of arrested metaphase blastomeres and an increase in nuclei with abnormal nuclear phenotypes such as anaphase bridging. These studies were the first to demonstrate that the spindle checkpoint is active in metaphase blastomeres and that a reduction in oxygen signals spindle checkpoint function. The specific signal from a reduction of oxygen to the activation of the spindle checkpoint apparatus is not worked out. However, there is a reduction in microtubule polymerization in anoxic metaphase blastomeres, suggesting that a decrease in microtubule structure may be the signal to the spindle checkpoint proteins to initiate an arrest of metaphase blastomeres (Hajeri et al., 2005). This is inline with the findings by others that drugs that perturb the microtubule structure lead to an induction in spindle checkpoint function. Since various spindle checkpoint alleles are associated with predisposition to some cancers, the importance the of spindle checkpoint function and oxygen levels in regards to human health related issues is further underscored (Hardwick et al., 1999; Hardwick and Murray, 1995; Hartwell, 2004).

4.3 Cellular changes and genes associated with prophase arrest

In comparison to a metaphase arrest, an arrest of a prophase blastomere is less characterized. To further analyze prophase arrest the progression of prophase to metaphase must be understood. In *C. elegans*, the transition from prophase to prometaphase occurs when the chromosomes are fully condensed and nuclear envelope break down (NEBD) begins (Dernburg, 2001; Moore et al., 1999; Oegema et al., 2001). The progression of NEBD, which is a commitment to mitosis, can be followed using cellular analysis to detect nucleoporins, which are components of the nuclear pore complex. In an anoxia-induced prophase arrested cell the process of NEBD and the transition to prometaphase is arrested. To further understand prophase arrest two main approaches have been taken. First, cell biological analysis of nuclear structures was conducted to characterize the prophase arrest. Second, RNAi screens and analysis of genetic mutants were conducted to identify genes required for anoxia-induced prophase arrest. These approaches are of interest to identify molecular changes in the arrested prophase blastomere and to identify genes essential for anoxia survival.

A hallmark of an anoxia-arrested prophase blastomere is that the condensed chromosomes associate with the inner nuclear periphery; we refer to this phenotype as "chromosome docking" (Table 2) (Hajeri et al., 2005). Interestingly, anoxia-induced chromosome docking occurs in both the somatic cells of the developing embryo and in the oocyte of an adult hermaphrodite exposed to anoxia (Hajeri et al., 2010). In the embryo exposed to anoxia the chromosomes will condense prior to movement to the inner nuclear periphery. The chromosomes will remain docked at the inner nuclear membrane until returned to a normoxic environment. This is in contrast to the normoxic embryo in which the chromosomes move throughout the nucleus until NEBD occurs. In anoxia-exposed adult hermaphrodites the oocytes, which are in prophase I of meiosis, contain bivalent condensed chromosomes that localize to the inner nuclear periphery. In contrast, the oocytes of normoxic controls contain bivalent chromosomes that localize throughout the nucleus (Hajeri et al., 2010). *Drosophila* embryos exposed to anoxia also contain chromosomes that associate with the inner nuclear periphery indicating that chromosome docking in response to anoxia is not just a *C. elegans* phenomenon (Foe and Alberts, 1985). The relevance of chromosome docking in blastomeres that are exposed to anoxia is unknown but it is possible that chromosome docking is a means to maintain chromosome integrity or function during anoxia exposure. While much is known about chromosomal territories in the interphase nucleus little is understood about chromosome location in prophase cells (Cremer et al., 2000; Geyer et al., 2011). It is not known if the mechanisms that regulate chromosome territories in interphase cells overlap with those regulating chromosome docking in arrested prophase blastomeres.

Given that anoxia induces chromosome docking in prophase blastomeres, indirect immunofluorescence has been used to characterize proteins associated with the nuclear membrane and chromosomes. Cell biological analysis shows that nuclear structures are altered in arrested prophase blastomeres relative to normoxic control embryos. Note that a relevant aspect of *C. elegans* chromosomes is that they are holocentric instead of monocentric in nature; this allows detailed cell biological analysis of chromosomes and chromosomal associated proteins. Using an antibody to recognize the kinetochore protein HCP-1 (CENP-F like) we determined that the kinetochore is altered in anoxia-arrested prophase blastomeres (Figure 2A). HCP-1 associates with chromosomes of normoxic prophase blastomeres but is not detected on the chromosomes of anoxia-arrested prophase blastomeres until embryos are returned to a normoxic environment (Figure 2A). In anoxia-arrested metaphase blastomeres HCP-1 is detectible indicating that the kinetochore changes observed in anoxia blastomeres is dependent on stage of mitosis (Hajeri, 2005).

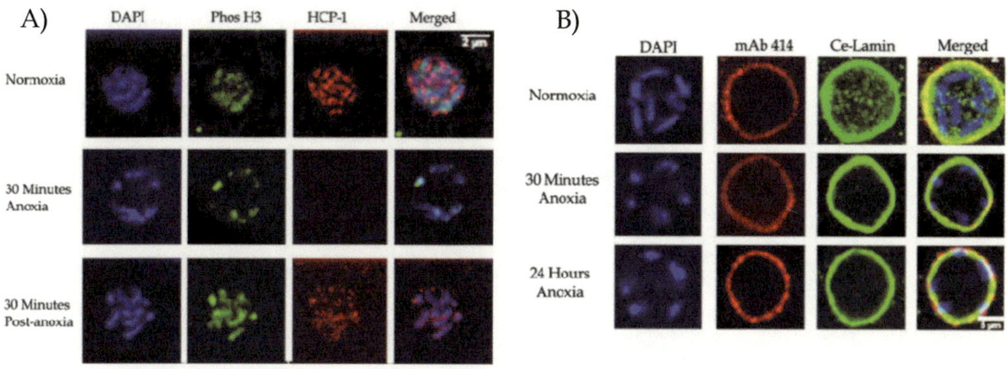

Fig. 2. Prophase blastomeres are altered in response to anoxia. A) Prophase blastomeres of embryos exposed to anoxia have diminished level of the kinetochore protein HCP-1. Embryos were collected from adult hermaphrodites and exposed to either normoxia or a brief period of anoxia and immediately fixed or allowed to recover in 30 minutes of air (post-anoxia) after anoxia treatment. After treatment the embryos were collected fixed and stained with DAPI to recognize DNA, mAb 414 to recognize the nuclear pore complex, anti phosphorylated Histone H3 at Serine 10 (Phos H3) to recognize mitotic nuclei and anti HCP-1 to recognize the kinetochore. Shown are representative prophase blastomeres analyzed using confocal microscopy. Scale bar = 2 µm. B) Lamin localization is diminished in the nucleoplasm of embryos exposed to anoxia. Embryos were exposed to normoxia or anoxia, for the specified time, and stained with DAPI to recognize DNA, mAb 414 to recognize nuclear pore complex and anti Ce-lamin to recognize lamin. Shown is a representative prophase blastomere from embryos exposed to noted environment and analyzed using confocal microscopy. Scale bar = 5µm.

Lamin, is an important inner nuclear protein that functions to maintain nuclear membrane structure and function. It is a target of post-translational modifications by CDK-1 during the complex process of cell cycle progression through mitosis (D'Angelo et al., 2006; De Souza et al., 2000; Gong et al., 2007; Heald and McKeon, 1990). In *C. elegans*, Ce-Lamin is localized to the inner nuclear membrane and nucleoplasm in normoxic prophase blastomeres. However, in the prophase blastomeres of anoxia-exposed embryos, Ce-Lamin is primarily localized to the inner nuclear membrane and there is a reduced level in the nucleoplasm (Figure 2B). The significance of reduced lamin in the nucleoplasm in anoxic blastomeres is not understood but does reflect alterations within the nucleus of anoxia-exposed embryos.

Antibodies that recognize nucleoporins associated with the nuclear pore complex can be used to monitor the nuclear envelop relative to cell cycle position (D'Angelo and Hetzer, 2008). We did not notice substantial change in the NPC of prophase-arrested blastomeres when assayed using mAb 414 (Table 2). However, using a commercially available antibody raised against human NUP50 we found evidence that an anoxia-arrested prophase blastomere differs in comparison to a normoxic prophase blastomere. The late prophase blastomeres of embryos exposed to normoxia have a reduced level of NPC that is detected by anti human NUP50, which is suggestive of NEBD occurring (Figure 3, arrow). Yet, in the anoxia-arrested prophase blastomere anti-NUP50 signal was present, suggesting that NEBD is not occurring and may be arrested (Figure 3, arrow head). Thus, a plausible mechanism to induce prophase arrest is to prevent NEBD and thus the transition to prometaphase. In both normoxic and anoxic embryos anti-NUP50 recognizes interphase nuclei in a similar manner.

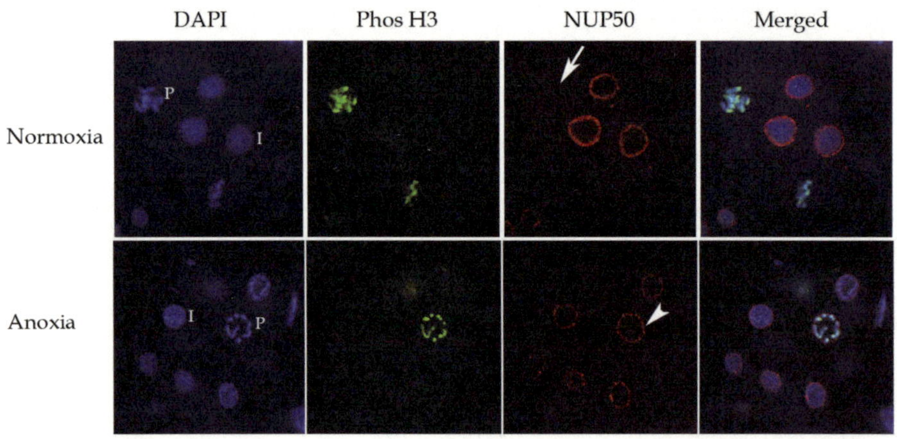

Fig. 3. Anti-NUP50 localizes to nuclear membrane and has a different pattern in anoxia exposed prophase blastomeres. The anti-human NUP50 antibody detects antigen localized to the nuclear membrane in interphase cells (I) which is then diminished by prophase (P, arrow) in normoxic embryos. In the prophase blastomeres (P) of anoxic embryos the antigen remained associated with the nuclear membrane (arrow head). Shown is a representative embryo analyzed by confocal microscopy.

Genetic analysis has been instrumental for identifying processes that regulate cell cycle arrest and progression. Previously, we determined that knockdown of the nucleoporin protein NPP-16/NUP50 by RNAi or genetic mutation results in a decrease in embryos that survive anoxia exposure (Hajeri et al., 2010). Additionally, in the anoxia exposed *npp-*

16(RNAi) embryo, there is an increase in abnormal nuclei and a reduction in arrested prophase blastomeres (Figure 4). The number of arrested metaphase blastomeres is not significantly different than wild-type embryos exposed to anoxia indicating that *npp-16* function is required specifically for prophase arrest (Hajeri et al., 2010).

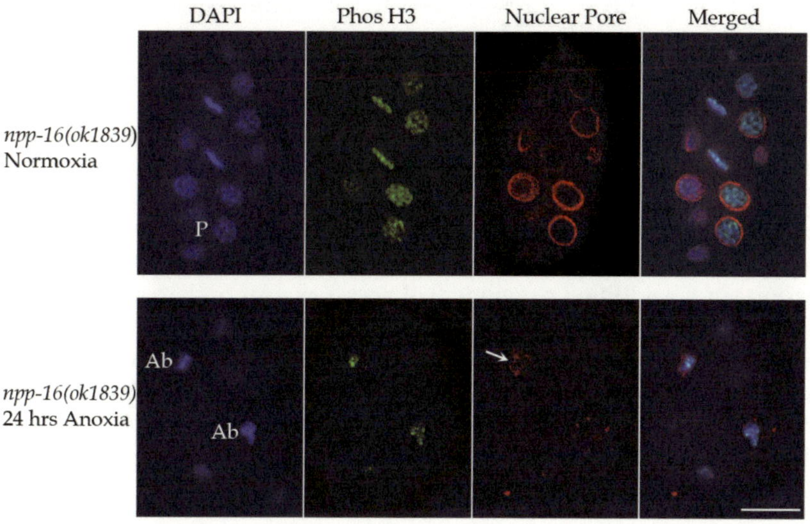

Fig. 4. The gene encoding the nucleoporin, *npp-16*/NUP50, is required for anoxia-induced prophase arrest. *npp-16(ok1839)* embryos were exposed to normoxia or anoxia and then stained with DAPI to detect DNA, Phos H3 to detect the mitotic marker phosophorylated Histone H3 at Serine 10, and mAb 414 to detect NPC. The *npp-16(ok1839)* embryos exposed to normoxia contain normal prophase (P) and yet the *npp-16(ok1839)* embryos exposed to anoxia contain a decrease in prophase blastomeres and an increase in abnormal nuclei (Ab) and NPC structure (arrow). Scale bar = 20 μm.

What is the role of NPP-16 in anoxia-induced prophase arrest? A key to addressing this question was noting that the predicted NPP-16 human homologue NUP50 was shown to interact with $p27^{kip1}$, a CDK inhibitor, suggesting a role of NUP50 with cell cycle regulation (Smitherman et al., 2000). In mammalian cells, CDK-1/cyclinB regulates the G2/M transition and NEBD by targeting a multitude of substrates (Lindqvist et al., 2009; Lindqvist et al., 2007). In *C. elegans* embryos, the NPC protein gp210, which is phosphorylated by CDK-1/cyclin B, is important for the depolymerization of lamin and required for NEBD (Galy et al., 2006). Data suggest that NPP-16 and CDK-1 have a role in anoxia-induced prophase arrest and that anoxia-induced arrest of NEBD is compromised in *npp-16* mutants. The use of antibodies that recognize CDK-1 showed that in wild-type embryos exposed to anoxia the protein is localized near the chromosomes in prophase blastomeres; this localization is reduced in the *npp-16* mutant exposed to anoxia. Second, an antibody that recognizes the inactive form of CDK-1 (anti CDK-1 P^{Tyr15}) was localized to prophase blastomeres of anoxic embryos but was absent from the prophase blastomeres of normoxic controls or the *npp-16* embryos exposed to anoxia. This indicates that not only is CDK-1 regulated differently in anoxic embryos but that this regulation differs in the *npp-16* mutant which does not arrest properly in response to anoxia. Although the specific signaling between NPP-16 and CDK-1 is not yet understood this work does provide evidence that anoxia influences cell cycle machinery.

Chromatin modifications have major affects on chromatin structure and function (Geyer et al., 2011). Modifications of histones are highly conserved in eukaryotes and influence many cellular processes such as gene expression and chromosome condensation. For example, the phosphorylation of histone H3 at Serine 10 is known to correlate with mitotic and condensed chromosomes. Previously, we showed that the phosphorylation of histone H3 at Serine 10 is reduced in mitotic blastomeres of embryos exposed to long-term anoxia. Alteration in the phosphorylated state of proteins may reflect that energy-requiring processes are reduced in anoxia and that cellular signals change in anoxia-exposed embryos. Acetylation of histones is another example of how post-translational modifications regulate cellular functions. Histone acetylation and deacetylation by Histone Acetyl Transferases (HATs) and Histone Deacetylase (HDAC), respectively, modulate chromatin and influence gene expression via the addition or removal of acetyl groups on histones (Ferrai et al., 2011). There are several *C. elegans* genes that are involved with histone modifications and many of these genes are essential. We found that the gene *hda-2,* when knocked down using RNAi, does not lead to embryo lethality or obvious defects in normoxic embryos. However, when these embryos are exposed to anoxia there is an increase in abnormal nuclei (Figure 5). The specific role *hda-2* has in anoxia response and survival in the embryo needs to be further analyzed.

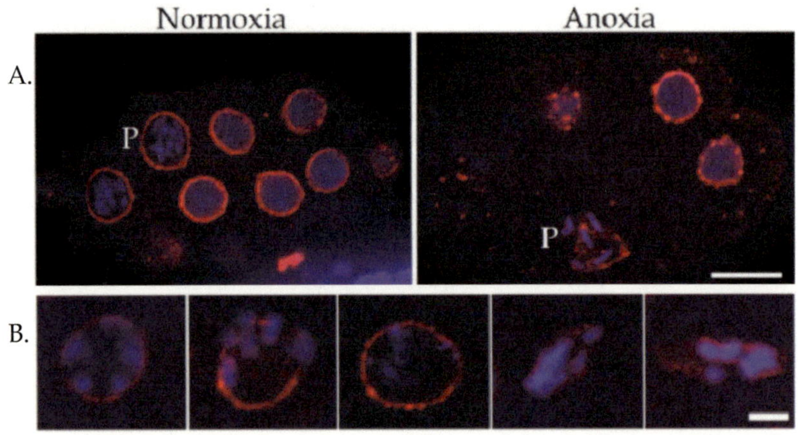

Fig. 5. *hda-2(RNAi)* embryos are sensitive to anoxia. Embryos were obtained from gravid adults and fixed and stained with DAPI to detect DNA and mAb 414 to detect the nuclear pore complex. A) Unlike normoxic controls, the *hda-2(RNAi)* embryos exposed to anoxia have abnormal blastomeres and prophase (P) blastomeres in which the chromosomes do not associate with the NE. Scale bar = 19 µm. B) Enlarged image of nuclei observed within the *hda-2(RNAi)* embryos exposed to anoxia. A significant number of the blastomeres contain a variety of abnormal nuclei. Scale bar = 5 µm.

There is evidence that modulation of the electron transport chain (ETC) activity has a role in cellular arrest. Exposure of embryos to ETC inhibitors (Ex: sodium azide) lead to cell cycle arrest and docking of prophase chromosomes. However, the embryos do not remain arrested and die within an hour of exposure (Hajeri et al., 2010). Thus, ETC inhibitors do not phenocopy anoxia exposure, suggesting that anoxia-induced suspended animation is partially regulated by changes in the ETC. Unanswered questions regarding anoxia-induced cell cycle arrest include: what is the specific signal between reduced oxygen levels and docking of prophase chromosomes? Is chromosome docking essential for anoxia survival? How are cell cycle checkpoint proteins regulated in the anoxia-exposed embryo? Further genetic analysis of *C. elegans* embryos exposed to anoxia can lead to answers to these questions.

4.4 Metabolic and environmental changes that influence anoxia survival in the embryo

In the embryo exposed to anoxia, not only does the cell cycle machinery need to respond to the reduction in oxygen levels but metabolic pathways must do so as well. Embryos exposed to anoxia have a reduction in the ratio of ATP/AMP; this reduction will affect metabolic pathways (Padilla et al., 2002). Given the central importance of carbohydrates to metabolism, carbohydrate homeostasis is likely to be important for anoxia survival. Indeed it was found that sugar levels are altered in anoxia exposed embryos. For example, glycogen levels decreased to approximately 20% of intitial levels after a 24 hour exposure to anoxia (Frazier and Roth, 2009). A sufficient level of carbohydrates, perhaps in the form of glycogen, is likely important for maintaining metabolism during anoxia exposure. The gene *gsy-1* encodes glycogen synthase and when this gene is knocked down by RNAi the animal has a reduction in glycogen stores are sensitive to anoxia. (Frazier and Roth, 2009). Mutations in other genes that have reduced glycogen levels were also sensitive to anoxia further supporting the idea that glycogen homeostasis is important for anoxia survival. The Frazier and Roth study also demonstrated that the environment to which the parent is exposed can influence the anoxia survival rate of its embryos. For example, when L4 larvae develop to gravid adhulthood in a high salt environment (300 mM sodium chloride) their embryos are sensitive to anoxia (Frazier and Roth, 2009). It is likely that alterations of other central metabolic macromolecules are important for anoxia survival and the use of *C. elegans* genetics to alter metabolic pathways will shed light on metabolic pathways required for anoxia survival.

Embryos are able to survive anoxia and hypoxia (0.5% O_2), yet when exposed to to severe hypoxia (100 to 1000 ppm of O_2) embryos will die (Nystul and Roth, 2004; Padilla et al., 2002). Whereas anoxia is inducing suspended animation a .5% O_2 hypoxic environment is sufficient to support the signal for developmental activities. It is possible that the embryos exposed to 100 to 1000 ppm of O_2 are exposed to oxygen levels that are too high to induce suspended animation but not high enough to support normal growth and development. In the initial experiments of embryos exposed to 100 to 1000 ppm of O_2 the embryos were on media and not within the adult (Nystul and Roth, 2004). However, when gravid adults were exposed to 100 to 1000 ppm of O_2 the embryos within the uterus survived by arresting (Miller and Roth, 2009). These results indicate that the O_2 microenvironment differs between the uterus of an adult and the surface of NGM media and that embryos differentially respond to O_2 levels.

5. Anoxia tolerance in adult animals

C. elegans adult animals have been useful for understanding the genetic and physiological responses to oxygen deprivation particularly because of the mechanistic overlap in oxygen deprivation responses between *C. elegans* and other metazoans including humans (Powell-Coffman, 2010). Several unique characteristics of adult *C. elegans* make it a valuable model to study responses to oxygen deprivation. First, the adult animal has a relatively simple anatomy, easily observable somatic tissues and meiotic cells. These tissues amenable to analysis include muscle, neurons, intestinal cells and a very well studied germline that contains meiotic cells that give rise to oocytes and sperm in the hermaphrodite. Second, *C. elegans* has been used by many within the community to study genes involved with stress responses and lifespan; these studies allow investigators to identify overlapping and distinct mechanisms between stress responses and lifespan. Finally, *C. elegans*, as a soil nematode, has adapted to changing oxygen levels in the environment. Taken together, the anatomical, genetic and environmental niche characteristics of *C. elegans* provide a unique opportunity to identify the ways in which this simple model responds to and survives oxygen deprivation. Such information can aid in our understanding of why species do or do not have limitations in oxygen deprivation response and survival.

Metazoans, including *C. elegans*, possess complex biochemical mechanisms that operate at the cellular level to promote oxygen deprivation tolerance (O'Farrell, 2001). These adaptations allow anaerobiosis in severe hypoxia and anoxia through a range of physiological responses that operate via three general strategies: increase the rate of flux through glycolytic pathways, decrease overall energy demand by rapid reduction in metabolic rate, or activation of physiological mechanisms that increase the efficiency of oxygen removal from the environment (Hochachka, 2000; Hochachka et al., 1996). The execution of these strategies involves modulation of a wide range of cellular pathways. For example, animals switch from energy source molecules during oxygen-deprivation from fat that is primarily utilized for aerobic energy metabolism to glycogen/glucose stores. During a 24 hour anoxia exposure as much as two-thirds of the animals carbohydrate reserve may be utilized as an energy source; this usage nearly accounts for the mass of glycosyl units of metabolites produced during the oxygen deprivation period (Foll et al., 1999).

C. elegans frequently encounters oxygen-deprived microenvironments in its natural habitat and adult animals have adapted to tolerate these exposures. Wild-type hermaphrodites that are 1-day old (one day after the L4 larval to adult molt) survive 24 hours of anoxia at ≥90% (20°C) when assayed on solid NGM medium (Padilla et al., 2002; Van Voorhies and Ward, 2000). Interestingly, Foll et al., (1999) reported a higher mortality for adult worms exposed to 24 hours of anoxia (20°C) and a subsequent sharp rise in mortality for slightly longer exposure when assayed in liquid culture. The discrepancy between the reported survival rates of the two studies is likely due to differences in methodology. One possibility is that the process of crawling across agar medium requires less energy than swimming in liquid medium. If so, the additional energy expenditure while swimming in liquid media may compromise anoxia tolerance. While adult hermaphrodites are anoxia-tolerant the survival rate plummets as the duration of anoxia is lengthened (Mendenhall et al., 2006; Mendenhall et al., 2009; Padilla et al., 2002). The 1-day old adult has a markedly decreased survival rate (4.7%) in long-term anoxia, defined as a 72 hour or more anoxia exposure at 20°C, demonstrating that there is an anoxia survival limitation (Mendenhall et al., 2006). The anoxia-survival limitation is taken advantage of to identify genetic mutations that lead to anoxia sensitivity (mutants that cannot survive 24 hours of anoxia) and anoxia tolerance (mutants that can survive long-term anoxia, > 3 days of anoxia).

The adult anoxia-tolerance strategies include the worm entering a reversible state of suspended animation. In this state adults do not feed, do not lay eggs and cease to be motile. The process of crawling has been reported to carry a relatively low metabolic cost to the worm compared to the high cost of reproduction and tissue maintenance and this assessment is supported by the observation that animals whose metabolic rate has been reduced by greater than 90% do not show abnormal motility (Van Voorhies and Ward, 2000; Vanfleteren and De Vreese, 1996). The length of time animals remain active after the onset of anoxia varies among *C. elegans* strains. The majority of wild-type adults cease movement within 8 hours of the onset of anoxia (Mendenhall et al., 2006). However, the *daf-2(e1370)* animal, which is a long-term anoxia tolerant mutant strain and carries a reduction-of-function mutation in the insulin-like receptor (see section 6 below), will delay entering into suspended animation as demonstrated by observable movement after 24 hours of anoxia. Although movement is observed in the *daf-2(e1370)* animal exposed to anoxia it is slower than normoxic controls (Mendenhall et al., 2006). To date no mutation has been isolated that prevents the worm from entering into a state of suspended animation.

The cylindrical body and simple gut design of the worm favors rapid diffusion of gases across both the gut lumen and cuticle into the metabolically active intestine. *C. elegans* is an oxygen regulator and seems to be insensitive to hyperoxia (Van Voorhies and Ward, 2000). However, when confronted with oxygen deprivation the worm must respond by either remaining animated or entering into suspended animation; the determining factor often being oxygen tension and perhaps metabolic state (Nystul and Roth, 2004). It has been observed that animals remain active in hypoxia but enter suspended animation in anoxia. Which factors are critical in the molecular decision to suspend or continue processes such as movement and how these factors are regulated at the cellular and tissue level remains unclear. Nevertheless, valuable information regarding genes required for both hypoxia and anoxia survival has been gleaned (Jiang et al., 2001; Padilla et al., 2002; Scott et al., 2002). For example, among the adaptations adults posses is the ability to sustain a steady metabolic rate even when exposed to a range of decreasing oxygen tensions and not until ambient oxygen tension falls to 3.6 kPa will metabolic rates begin to drop for young adults (Anderson and Dusenbery, 1977; Suda et al., 2005; Van Voorhies and Ward, 2000). However, once the environment becomes anoxic, metabolic rate drops to as low as 5% of that in normoxic conditions and recover in a slow linear fashion only after removal from anoxia (Van Voorhies and Ward, 2000).

5.1 The anatomical and physiological impact of anoxia exposure

While in a state of suspended animation, the immobile *C. elegans* often adopt linearly extended bodies or a bent or curved-sickle shape (Figure 6, arrow). Upon re-oxygenation survivors will resume movement and the overwhelming majority exposed to 24 hours of anoxia will move normally several hours post recovery (Figure 6B). Initial movement begins with slight side-to-side movement of the anterior head region then slowly spreads to include the entire head region. As recovery progresses the worm regains the ability to move the mid-body and tail regions in the classic sinusoidal motion and resumes foraging and egg-laying. Recovery from long-term anoxia takes longer and not all physiological processes appear to resume at the same rate. For example, in the few wild-type animals that survive long-term anoxia, contraction of the somatic gonad sheath and ovulation has been observed within 12 hours of post-anoxia, which is often before full body motility has resumed.

Recovery of anatomical and organ function at different rates may compromise the viability of the animal. For example, if ovulation precedes the ability to lay eggs, the accumulated eggs within the uterus may lead to embryos hatching within the uterus (bagging out phenotype) and further compromise organs such as neurons, muscles or the intestine. It is possible that long-term anoxia survivors are better able to resume anatomical and physiological processes.

The extent to which animals regain motility after anoxia exposure can vary within an experimental cohort and is influenced by duration of anoxia and genotype (see section 6). As the duration of anoxia exposure is lengthened the number of individuals that regain normal motility decreases (as visualized via a dissection microscope). In regards to wild-type adult animals, recovery within minutes can be observed after a 24-hour anoxia exposure. However, the few wild-type animals that survive long-term anoxia may not begin moving for several hours after re-oxygenation. This variability has been a useful tool in assessing the post-anoxia condition of survivors. Several approaches have been described for categorizing worm locomotion phenotypes (Gerstbrein et al., 2005; Herndon et al., 2002; LaRue and Padilla, 2011). In each system animals are categorized based on their movement and response to touch. For example, an animal is scored as dead if it is not moving and does not respond to gentle touch with a platinum wire. If the animal moves in response to touch it is scored as alive and then further classified based on level of motility; this classification provides an assessment of how well the animal moves after anoxia exposure. That is, animals capable of completing the typical sine wave motion similar to that of untreated adults are classified as having "unimpaired" movement while animals that move abnormally or move only a portion of the body are classified as having "impaired" movement. Utilizing this method one can assess how well an animal tolerates anoxia by monitoring its ability to recover and execute the fundamental process of movement. Often the impaired worms do not move and will consume the bacteria nearby leaving a fan-shaped area emptied of food (Figure 6D). The underlying cause (Ex: a compromise of muscle and/or neuronal function) of anoxia-induced impairment is yet to be determined.

It is also possible that post-reoxygenation impairment is due, at least in part, to the organism's inability to execute the processes required for the maintenance of cellular integrity thus leading to a loss of tissue structure (Mendenhall et al., 2006). Wild-type animals recovering from long-term anoxia have an overall loss of tissue structure in the head region that contains both neuron and muscle. Along with distortions in the muscle isthmuses of the pharynx, relatively large vacuoles or cavities also appear throughout the soma (Figure 7). However, long-term anoxia tolerant strains do not show the same tissue disorganization and appear to be better able to maintain tissue structure. This is presumably accomplished by either sustaining a homeostatic physiology during the anoxic period or by activating tissue maintenance and repair pathways post-reoxygenation.

In addition to the loss of coordinated movement and incurring tissue damage, anoxia exposure affects multiple aspects of fertility. When eggs are exposed to 24 hours of anoxia and then allowed to mature to adulthood the onset of first reproduction is significantly delayed compared to normoxic controls. The anoxia treated animals also have a reduced reproduction rate and reduced fecundity compared to untreated controls (Van Voorhies and Ward, 2000). It is possible these changes are the result of a programmed response to the stress or may simply be the result of damaged meiotic cells.

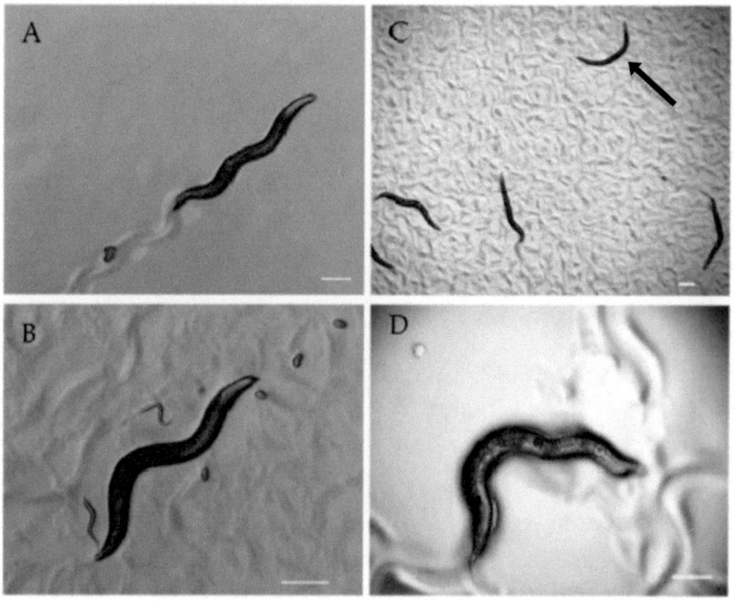

Fig. 6. Wild-type *C. elegans* adults survive and fully recover motility after 24 hours of anoxia whereas animals exposed to 72 hours of anoxia have a reduced survival rate and impaired motility. Wild-type animals were raised to one-day old adults then exposed to anoxia for 24 hours or 72 hours. A) One-day old adult hermaphrodite prior to anoxia exposure. B) The same adult following 24 hours of anoxia and given 1-hour of recovery in normoxia. Animal recovered normal pattern of movement and resumed egg-laying within an hour of re-oxygenation. C) One-day old adult hermaphrodites in suspended animation immediately following 72 hours of anoxia. Note the slightly curved body posture (arrow). D) Example of an impaired survivor following 72 hours of anoxia and 24 hours recovery in normoxia. Note the impaired animal has consumed the bacterial food in a fan-shaped halo surrounding the anterior head region. All anoxia exposures were conducted at 20˚C. Scale bar = 100 μm (A, B, D); scale bar = 20 μm (C).

5.2 Environmental factors affect anoxia tolerance

The environment to which an organism is exposed can have profound affects on phenotype. Environmental changes that induce a stress response include changes in temperature, water availability, diet and nutrient content, oxygen levels and exogenous molecules including toxins or pharmaceutical agents. If an organism is exposed to these factors during development and/or adulthood, the responses to and ability to survive the stress can be affected. In some cases a preconditioning environment, a low level of stress, can improve stress tolerance.

In the laboratory *C. elegans* are typically grown at 15-20°C and provided the *E. coli* OP50 strain as a food source. There is evidence that the environment in which the animal is exposed will precondition for an enhanced anoxia survival phenotype by altering the physiology of the animal (LaRue and Padilla, 2011). Wild-type *C. elegans* grown at 20°C and fed the *E. coli* OP50 diet are very sensitive to long-term anoxia in that the majority of the

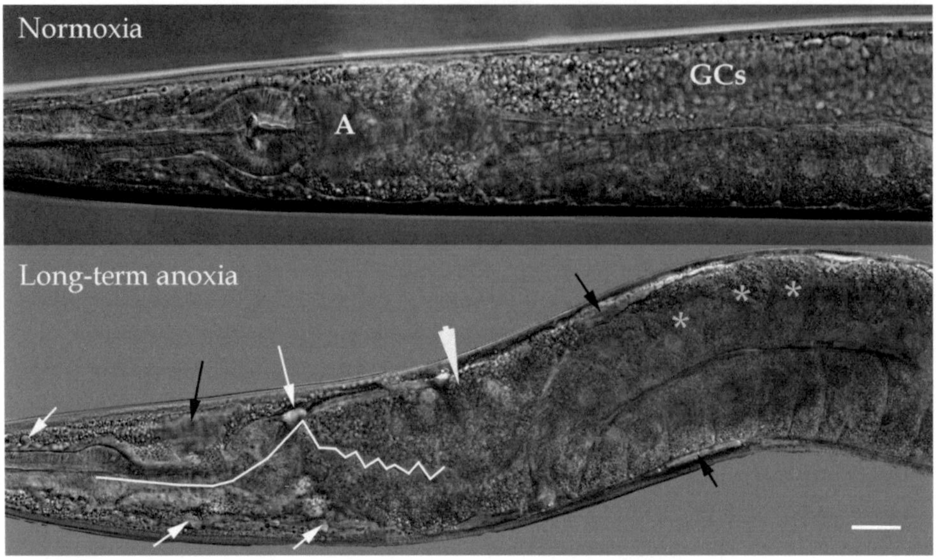

Fig. 7. Long-term anoxia exposure results in tissue abnormalities. One-day old adult wild-type hermaphrodites were exposed to 72 hours of either normoxia or anoxia followed by 24 hours of recovery in normoxia at 20°C. Anoxia treated animals, in comparison to controls, have an overall more grainy appearance, cavities/vacuoles (white arrows) in the psuedocoloem and head region. In addition, the anoxia-exposed animal shows accumulated fluid (black arrows) around the gut, intestine and pharynx. The unexposed animal has a normally structured intestinal lumen, which forms a large atrium-like cavity (A) at the anterior end of the gut at the pharynx-lumen juncture. In contrast, the anoxia survivor has bends in the pharynx and an intestinal lumen that is constricted and distorted forming irregular jagged kinks. The white line traces the lumen of the gut from the anterior bulb of the pharynx through the first intestinal cell. The intestinal cells of the anoxia-exposed worm are also packed with many very small intracellular globules (white arrowhead). The germline morphology is also affected by the anoxia stress. Asterisks mark the nucleus of some of the oocytes which are abnormally stacked well beyond the gonad bend in the anoxia treated animal compared to presence of syncytial germ cells (labeled GCs) visible in the distal gonad of the control animal. Images are both composites of three individual frames reassembled using Adobe PhotoShop CS. Scale bar = 20 μm.

animals die and among those that survive most have an impaired phenotype. However, if the animal is grown at 25°C and fed the *E. coli* HT115 strain throughout larval development there is a significant increase in long-term anoxia survival and an unimpaired phenotype. The animals grown at 25°C and fed the *E. coli* OP50 strain also survive long-term anoxia yet many have an impaired phenotype in that they display visible defects in motility and tissue morphology. These data suggest that growth at 25°C and a diet of *E. coli* HT115 during development may synergistically enhance anoxia survival for *C. elegans*. It is possible that the 25°C temperature induces stress response genes (Ex: heat shock proteins) that prepare the animal to survive long-term anoxia. Alternatively, the preconditioning environment could alter energy stores leading to an increase in anoxia survival. There is evidence to support the idea that metabolic stores are altered in *C. elegans* raised at 25°C and fed the *E. coli* HT115 diet during development.

First, it is known that the *E. coli* HT115 strain has higher carbohydrate levels in comparison to the OP50 strain; this may influence the metabolism of the worm (Brooks et al., 2009). Second, staining with carminic acid, which is used to detect carbohydrate stores within the intestines, indicates that carbohydrate levels are increased in the intestine of animals grown at 25°C and fed the *E. coli* HT115 diet in comparison to those raised at 20°C (LaRue and Padilla, 2011). Further analysis is needed to determine mechanistically how preconditioned metabolic and physiological changes within the nematode contribute to the enhanced long-term anoxia phenotype.

While the temperature 25°C is not typically considered a stressful environment for *C. elegans*, some physiological processes are likely different between animals grown at 25°C or greater in comparison to those grown at 15-20°C. An increased temperature such as >28°C is likely stressful to the animal. It is known that animals exposed to one day of anoxia or severe hypoxia at 28°C instead of 20°C leads to markedly decreased survival rate (Mendenhall et al., 2006; Scott et al., 2002). Animals that are exposed to more than one stress at a time typically have a decrease in survival. However, such a condition has been useful in identifying genetic changes that increase viability in this environment. Alleles that affect the insulin signaling pathway increase viability when exposed to anoxia and 28°C (Scott et al., 2002). The insulin signaling pathway is known to be important for many stress responses and it will be of interest to tease apart the molecular factors that are specific to anoxia response and survival.

6. Genetic factors associated with anoxia survival in adult animals

A major strength of the *C. elegans* model system is the ability to dissect biological processes using a genetic approach. The use of forward genetic screens, RNAi genomic screens, suppression or enhancer screens, and analysis of transgenic fusion proteins is fundamental for discovering mechanisms regulating biological processes. These approaches have been used by many within the *C. elegans* community to unravel the mechanisms regulating complex processes such as programmed cell death, aspects of embryo and larvae development, chemosensing, dauer development, meiosis and many other processes. *C. elegans* is also a wonderful model to study stress responses, environmental and genetic factors that influence lifespan and the overlap between mechanisms regulating stress responses and lifespan. In terms of the biology of the adult hermaphrodite, it contains meiotic cells that can differentiate into oocytes and sperm. The hermaphrodite, which produces approximately 300 offspring, typically lays the majority of the embryos within the first two days of adulthood. with offspring production tapering off after the third day of adulthood. Males exist within the population thus genetic crosses can be conducted. In typical laboratory conditions the adult hermaphrodite has an average lifespan of approximately 15-20 days. The capability to manipulate genotypes, a rapid lifecycle and capacity to produce a large number of offspring make *C. elegans* an excellent system to analyze stress responses in relationship to specific biological function, such as germline and metabolic capacity. Here we expand upon the genetic approaches used to analyze anoxia responses in adult *C. elegans*. Table 3 summarizes genotypes discussed further in this chapter and their respective phenotypes relative to anoxia tolerance, germline function and lifespan.

Genotype	Summary of Reproductive Capability and Lifespan
Long-term anoxia tolerant strains: high survivor rate and unimpaired phenotype after three days of anoxia and recovery	
daf-2(e1370)	Enhanced anoxia tolerance (5 days of anoxia); Brood size is slightly less than wild-type; intermittent progeny production late in life; lifespan is extended relative to wild-type
fem-3(q20)	Complete sterility; mutant produces sperm but no oocytes; wild-type lifespan
fog-2(q71)	Mutant is female and rarely lays unfertilized oocytes; male has functional sperm; wild-type lifespan
glp-1(e2141)	Complete sterility; somatic gonad present; primordial germ stem cell production abnormal; mutants have extended lifespan
ksr-1(ku86)	Reduction in meiotic progression and reduced progeny production
wild-type males	Have functional sperm; mean lifespan of 16 days is density dependent
wild-type fed FUDR	Exposure of hermaphrodites to FUDR at L4 stage induces sterility
Anoxia survival rate similar to wild-type: High survivor rate and unimpaired phenotype after 24 hours of anoxia but long-term anoxia sensitive	
wild-type hermaphrodite	Produce both oocytes and sperm; mean lifespan approximately 20 days
daf-2(e1370);daf-16(mu86)	Null *daf-16* allele suppresses *daf-2(e1370)* longevity and long-term anoxia tolerance
daf-2(e1370);gpd-2/3(RNAi)	*gpd-2/3(RNAi)* suppresses *daf-2(e1370)* high temperature and anoxia tolerance
daf-16(mu86)	Wild-type brood size; mean lifespan approximately 15 days
hif-1(RNAi)	Anoxia survival rate comparable to wild-type, sensitive to hypoxia
Anoxia sensitive: low survival rate and/or impaired phenotype after 24 hours of anoxia	
aak-2(rr48)	Survives anoxia but highly impaired after exposure; wild-type lifespan
gpd-2/3(RNAi)	Survives 24 hour of anoxia well; arrest motility immediately upon encountering anoxia

Table 3. Summary of anoxia tolerance, reproductive capacity and lifespan for specific genotypes discussed in this chapter.

6.1 Reduction in insulin-like signaling favors anoxia tolerance

Arguably the most well known pathway in the study of anoxia tolerance in *C. elegans* is the insulin/IGF-1-mediated signaling (IIS) pathway. Research led to identification and characterization of the molecular nature of the genes functioning in the insulin-like signaling pathway revealing that the pathway regulates metabolism, lifespan, stress responses and the development of the dauer state, which is a stress-resistant larva in diapause (Gottlieb and Ruvkun, 1994; Kenyon et al., 1993; Kimura et al., 1997; Riddle et al., 1981; Tissenbaum and Ruvkun, 1998). Much is known about the genes that function in *da*uer *f*ormation pathway (*daf*). The dauer regulatory pathway involves the *daf-2* and *daf-16* genes, which encode the insulin/IGF-1 receptor-like protein and a fork-head transcription factor, respectively (Kimura et al., 1997; Larsen, 2001; Larsen et al., 1995; Lin et al., 1997; Riddle et al., 1981). It is thought that DAF-2 interacts with a variety of insulin-like ligands and sends a signal via the AGE-1/PI3/AKT signaling pathway to repress the translocation of the transcription factor DAF-16 into the nucleus (Kenyon, 2010). The activation of functional DAF-2 results in phosphorylation of cytosolic DAF-16, an action that prevents its translocation into the nucleus keeping it sequestered in the cytoplasm. However, when signaling through the IIS pathway is reduced, for example during periods of food deprivation or in the presence of null or severely reduced function forms of the DAF-2 receptor, this inhibition does not occur and DAF-16 is translocated into the nucleus of the cell where it is thought to link with other nuclear factors to induce expression of a variety of genes in a coordinated manner to promote dauer formation, longevity, fat metabolism, stress response, innate immunity and anoxia tolerance (Kenyon et al., 1993; McElwee et al., 2003; Mendenhall et al., 2006; Murphy et al., 2003; Oliveira et al., 2009).

The *daf-2(e1370)* mutant allele confers a greatly extended lifespan (from 18 to 42 days) when worms are grown early in development at a permissive temperature (functional DAF-2 is present) then shifted to the non-permissive temperature (non-functional DAF-2) at the L4 stage of development or when grown continually through development at 20°C (Kenyon et al., 1993). In addition to modulating lifespan, the *daf-2(e1370)* allele also confers various stress responses including long-term anoxia tolerance (Mendenhall et al., 2006). Both exceptional phenotypes are DAF-16 dependent and animals carrying a null mutation allele of *daf*-16 show no extension of lifespan and are long-term anoxia sensitive (survival = ~2%). It is noteworthy that factors influencing stress response and lifespan have both common and distinct genetic signals. Further investigation of the overlap in these pathways is of interest to the study of anoxia response and tolerance.

6.2 Metabolic regulation is linked to anoxia tolerance

Several *daf-2* alleles induce a long-term anoxia or high-temperature anoxia/hypoxia survival phenotype; these phenotypes are suppressed by mutations in *daf-16* (Mendenhall et al., 2006; Scott et al., 2002). An RNAi screen of genes known to be up-regulated by DAF-16 led to the identification of the *gpd-2* and *gpd-3* genes; these genes are nearly identical at the amino acid level and encode two of four glycolytic enzyme isoforms of glyceraldehyde-3 phosphate dehydrogenase (GAPDH; GPD-2/3) (Mendenhall et al., 2006). The *daf-2(e1370);gpd-2/3(RNAi)* animal exposed to one day of anoxia (28°C) or long-term anoxia (20°C) has a significantly reduced viability in comparison to *daf-2(e1370)* animals. The *gpd-2/3(RNAi)* animals survive short-term anoxia exposure yet are often impaired. These observations demonstrate that the physiological state generated by the *daf-2(e1370)* mutation is capable of protecting somatic tissue during anoxic stress and that *gpd-2* and *gpd-3* suppresses the *daf-2(e1370)* anoxia tolerance phenotype. Other genes involved with glycolysis were knocked down by RNAi but did not result in an anoxia sensitivity phenotype suggesting that the anoxia sensitive phenotype due to knockdown of *gpd-2/3* may be due to something other than changes in glycolysis (Mendenhall et al., 2006).

The ability to survive periods of environmental stress such as anoxia involves integration of signals emanating from many sources. The extent to which adaptive response programs are activated should correspond to the level or intensity of the encountered stresses. Transcription and translation are modulated to decrease production of unnecessary gene products while ensuring proper levels of immediately necessary ones. Execution of the appropriate pathways and processes require adequate accessibility to energy, specifically ATP. 5′-AMP-activated protein kinase (AMPK) is one of the energy sensors that monitors cellular AMP/ATP ratios and is conserved between humans and nematodes (Beale, 2008). In even small decreases in cellular energy status, AMPK will operate on substrates such that anabolic pathways are stimulated and catabolic ones inhibited. Stress triggers of AMPK activation include glucose deprivation, ischemia, oxygen deprivation, exercise and skeletal muscle contraction. However, the key-activating trigger for AMPK is probably starvation making its primary role to function as a whole body energy balancer (Hardie et al., 2006). LaRue and Padilla (2011) evaluated the role of AMPK in anoxia tolerance. While the overall survival rate of wild-type hermaphrodites and *daf-2(e1370)* were not affected by knockdown of *aak*-2 compared to untreated controls there was a significant decrease in the number of animals surviving in an unimpaired condition. However, after 4 days of anoxia *aak-2(RNAi)* suppressed the survival rate and unimpaired phenotype in both wild-type animals grown at 25°C and *daf-2(e1370)* animal (LaRue and Padilla, 2011). These observations implicate AMPK as a player in anoxia tolerance and necessary for preventing loss of coordination during anoxia treatment.

Through work with other metazoan species it has been shown that during periods of anoxia a significant rise in the activities of enzymes responsible for glycogen degradation occurs in liver (Mehrani and Storey, 1995). *C. elegans'* simple body design localizes many of the functions accomplished by a variety of organs in higher eurkaryotes almost exclusively to the intestine, including carbohydrate storage (McGhee, 2007). In the long-term anoxia tolerant mutant strain *daf-2(e1370)*, metabolism favors production of fat and glycogen in the intestine and hypodermal cells (Kimura et al., 1997). LaRue and Padilla (2011) used carminic acid to investigate the effect of anoxia on levels of stored carbohydrates in wild-type and long-term anoxia tolerant strains including *daf-2(e1370)*. Carminic acid is a fluorescent derivative of glucose that binds to glycogen and trehalose. As expected, animals exposed to long-term anoxia showed a decrease in carminic acid staining post anoxia supporting the assumption that carbohydrates stores are utilized as an energy fuel during anoxic stress. They determined that wild-type adults grown at 25°C had higher levels of carminic staining in the intestine than control animals grown at 20°C and significantly elevated survival rates when exposed to 3 or 4 days of anoxia relative to 20°C controls. The long-term anoxia tolerant strains *daf-2(e1370)* and *glp-1(e2141)* both had high levels of carminic acid staining prior to anoxia exposure. RNAi knockdown of *aak-2* suppressed this high level of staining indicating a reduction in stored carbohydrate levels. When *daf-2(e1370);aak-2(RNAi)* animals were exposed to 3 days of anoxia they showed impaired motility compared to *daf-2(e1370)* controls. Furthermore, *aak-2* knockdown suppressed the *daf-2(e1370)* long-term anoxia tolerant phenotype when exposed to extended anoxic stress (4 days). Together these observations suggest that the level of carbohydrate available to the worm for use as fuel at the time it encounters anoxia can influence its ability to tolerate the stress, and that preconditioning at 25°C may operate at least in part by increasing the amount of stored carbohydrate available during anaerobiosis.

Interestingly, AMPK activity has also been implicated as the master metabolic regulator of lifespan extension in *C. elegans*, particularly under starvation conditions. There is evidence that *aak-2* promotes lifespan extension in the IIS mutants such as *daf-2* in a *daf-16*-independent manner (Apfeld et al., 2004). AMP/ATP ratios do not differ between wild-type and *daf-2*, suggesting that the longevity phenotype of *daf-2* mutants is not simply due to an altered ratio of the two nucleotide molecules. Furthermore, individuals with the *daf-16(mu86);aak-2(ok524)* genotype have a reduced lifespan compared to wild-type or individuals carrying each mutation separately. While the long-term anoxia tolerant phenotype of *daf-2(e1370)* is completely suppressed by loss of *daf-16*, loss of *aak-2* does not reduce the overall survival rate but instead significantly affects post-anoxia health. Taken together these observations suggest the genes function to influence lifespan and anoxia-tolerance phenotypes via separate pathways. It will be of interest to determine how other metabolic mutants, such as *daf-16(mu86);aak-2(ok524)* fares in long-term anoxia.

It is possible that alternative forms of carbohydrates naturally present in *C. elegans* may play a role in the extreme phenotypes of longevity and long-term anoxia tolerance. For example, trehalose is a glucose disaccharide that is thought to participate in a wide variety of stresses including heat, desiccation, hypoxia, oxidative stress and others. It has been proposed that trehalose exerts its stress-protective effects through protein stabilization (Hottiger et al., 1994; Singer and Lindquist, 1998). Lifespans of young-adult animals were optimally extended (by 32%) when the animal was exposed to 5mM trehalose but not by other disaccharides (Honda et al., 2010). A decrease in age-associated decline was seen within a few days of initial exposure to the sugar and the lifespan extension effect was greater in older animals than younger. Furthermore, trehalose-treated animals had an extended reproductive span that was not due to reduced daily progeny production but by prolonged self-fertility. Animals fed trehalose also showed other evidences of slowed aging and senescence, including delay of age-associated decline in pharyngeal pumping and reduced rate of accumulation of age-pigment. Interestingly, *Drosophila* adults overexpressing *tps-1*, trehalose-phosphate synthase, and with a confirmed increase in trehalose production had a reduced recovery time following anoxia exposure (Chen and Haddad, 2004). Furthermore, they present evidence and an argument supporting the role of trehalose as a protein stabilizer that operates during stress such as anoxia. The role of trehalose in the anoxia tolerance of *C. elegans* has not yet been clearly established. Considering the importance of metabolic factors in anoxia tolerance, it would be of interest to determine if trehalose plays a role in establishing the anoxia-tolerant phenotype.

6.3 Loss of ceramide signaling confers hypersensitivity to anoxia

The alleles identified that influence anoxia tolerance are mutations that lead to an increase in anoxia survival and were previously shown to influence stress responses, germline function or lifespan. Recently, a mutation in the *hyl-2* gene was isolated that leads to anoxia sensitivity in the adult hermaphrodite (Menuz et al., 2009). In carrying out a genetic screen, the researchers specifically sought mutations that suppressed 24 hour anoxia-survival at 20°C; this led to the identification of the *hyl-2(gnv1)* allele. The *hyl-2* gene encodes one of three ceramide synthases and has homology to Lag1p (yeast longevity assurance gene). Two alleles of a related ceramide synthase, *hyl-1(gk203)* and *hyl-1(ok976)*, carry loss of function mutations. In contrast to *hyl-2(gnv1)* the two loss of function alleles actually conferred an increased tolerance to 48 hours and 72 hours of anoxia. The HYL-1 and HYL-2 synthases operate to efficiently produce ceramides and sphingomyelins of different lengths. Presence of a functional *hyl-1* gene is not sufficient to rescue the anoxia sensitive phenotype of *hyl-2* deficient worms. This suggests that *hyl-2* operates to synthesize a specific ceramide required for anoxia tolerance. This is supported by the observation that the *daf-2(e1370);hyl-2(gnv1)* double mutant has a reduced anoxia survival compared to the *daf-2(e1370)* further suggesting that *hyl-1* and *hyl-2* work in parallel to affect anoxia tolerance. Ceramides have been implicated as effectors of several biological processes and it is possible that chemical interactions between ceramides of a specific chemical composition and other molecules may result in regulation of pathways specific to anoxia tolerance. It will be useful to clarify the role of ceramide-signaling in oxygen-deprivation tolerance as an approach to understanding the function of small lipophilic molecules in the regulation of biological processes.

6.4 The germline influences anoxia tolerance

As 1-day old adults, wild-type hermaphrodites actively reproduce via self-fertilization. Through the process of gonadal sheath contraction and dilation of the spermatheca mature oocytes move into the spermatheca to complete fertilization and are ovulated into the uterus theoretically making room for the next proximal maturing oocyte to take its place. These steps are initiated by binding of MSP (major sperm protein) to surface receptors on the proximal oocyte (Greenstein, 2005; Miller et al., 2001). Adult hermaphrodites undergoing oocyte maturation, fertilization and ovulation do not survive long-term anoxia (Mendenhall et al., 2009). In contrast, sterile animals that do not undergo oocyte maturation and ovulation (ex: *glp-1(e2141)*, *fog-2(q71)* and *fem-3(q20)*) and animals with reduced progeny due to a reduced rate of ovulation (ex: *ksr-1(ku68)*) display long-term anoxia tolerant phenotype that is *daf-16* independent.

The *glp-1* gene encodes an N-glycosylated transmembrane receptor that is one of two members of the LIN-12/Notch family of receptors present in *C. elegans*. Loss of function mutations of *glp-1* gene cause germ cells, located in the distal gonad that would normally undergo mitosis, to prematurely enter meiosis thus preventing formation of self-renewing germ cells and a functional germ line. Therefore, while *glp-1(e2141)* sterile mutants have a somatic gonad they are incapable of producing oocytes and sperm (Crittenden et al., 1994; Mendenhall et al., 2009). Anoxia survival analysis of 1-day old adult *glp-1(e2141)* hermaphrodites showed them to be long-term anoxia tolerant with a survival rate of approximately 98% (Mendenhall et al., 2009). LaRue and Padilla (2011) were able to partially suppress the *glp-1(e2141)* long-term anoxia tolerant phenotype when the *aak*-2 was knocked down via RNAi in the double mutant *glp-1(e2141);daf-16(mu86)*. It is worth noting that in addition to having an anoxia-tolerant phenotype, sterile *glp-1(e2141)* mutants also have an increased lifespan relative to wild-type animals (Arantes-Oliveira et al., 2002).

Sterile genetic strains may exist as temperature-sensitive genetic mutants such as the germ-line deficient mutant strain *glp-1*(*e2141*), or as gonochoristic mutant strains such as *fog-2*(*q71*) in which females are incapable of producing self-sperm and thus the strain is maintained by mating with males. Additionally, treatments such as feeding animals the cell-cycle inhibitor drug FUDR or laser ablation of the germline precursor cells in L1 larvae will result in sterile animals. There is not only a relationship between sterility and anoxia survival but sterility also has an influence on increased lifespan. Interestingly, the longevity phenotype of sterile mutants is not due to merely the absence of producing offspring. Laser ablation of the germline precursor cells results in animals without a germline yet still possessing an apparently fully developed somatic gonad; such animals show the lifespan extension phenotype. However, ablation of both the germline and somatic gonad precursor cells results in sterile adults with a wild-type lifespan. Since both ablation treatments render the worm sterile the difference in lifespan cannot be attributed just to reproductive cost. Instead these studies present substantial evidence that the somatic gonad and germline both influence lifespan in contrasting manner (Hsin and Kenyon, 1999; Kenyon, 2010). While absence of a germline in *glp-1(e2141)* results in an long-term anoxia tolerant phenotype the role, if any, played by the somatic gonad in anoxia tolerance has not been determined.

The anoxia-tolerant phenotype of the unmated *fog-2(q71)* is suppressed by mating with a fertile male. This observation supports the theory that the maternal soma is under the regulatory control of the germline. While the mechanism by which the germ line regulates maternal log-term anoxia sensitivity is not yet known, exceptions to the observation that sterility induces long-term tolerance are known. First, mutant strains have been identified that are sterile yet long-term anoxia sensitive. The sterile strains *spe-9(hc52ts)* and *fer-15(hc15)* are capable of completing the initial steps of oocyte maturation but produce no viable offspring, yet both strains are long-term anoxia sensitive (survival rate= 23.4% and 2.6%, respectively). This sensitivity is presumably due to an altered physiology triggered in the somatic tissues in response to signals originating in the germline. Second, in a contrasting exception, the mutant strain *daf-2(e1370)* is not only long-term tolerant but also fertile (Larsen et al., 1995). At 15°C *daf-2*(*e1370*) has a slightly smaller brood size than wild-type (81% of wild-type). Furthermore, the *daf-2*(*e1370*) animals lays eggs over an extended period of adulthood (from adult day 1- 6) compared to wild-type (from adult day 1- 4) (Larsen et al., 1995; Tissenbaum and Ruvkun, 1998). It is unlikely that a reduction in average daily progeny production alone is sufficient to account for the strong long-term anoxia tolerant phenotype seen in a 1 day old adult *daf-2*(*e1370*). Instead, the reduction in function of DAF-2 is probably operating by acting on substrates and in pathways not yet identified. Finally, It is important to note that unlike *daf-2(e1370)* the anoxia-tolerant phenotype of these sterile reproductive mutants is *daf-16*-independent. Evidence thus far suggests that the long-term anoxia tolerant phenotype can be established via multiple pathways that may genetically overlap but which are not identical.

6.5 Anoxia tolerance is sex influenced

The overwhelming majority of stress response studies, at the genetic and cellular level, have been conducted in adult hermaphrodites. This is likely due to the ease in obtaining and maintaining hermaphrodite animals in comparison to males. Yet, analysis of males and their response to stress may provide insight into the understanding of mechanistic responses to and survival of anoxia. The wild-type male and hermaphrodite differ in several respects aside from the obvious sex-differentiated phenotypes such as different germline structure and function. For example, the lifespan of males is shorter than that of hermaphrodites and male lifespan is dependent upon whether the individual male is solitary or within a group of other males (Gems and Riddle, 2000). Gems and Riddle interestingly found that males that are solitary have a longer lifespan than males that are cultured as a group of other males indicating that male-male interactions reduce lifespan.

Survival of long-term anoxia also differs between wild-type adult hermaphrodites and males. One-day old wild-type and *daf-16(mu86)* mutant hermaphrodites survive 72 hours of anoxia at approximately 10% and 7% respectively, and are considered long-term anoxia sensitive. In contrast, wild-type and *daf-16(mu86)* mutant males survive long-term anoxia with a viability >98% (Mendenhall et al., 2009). The animals maintain normal motility and demonstrate an unimpaired phenotype after long-term anoxia exposure. Furthermore, the males that were raised in the presence of hermaphrodites and likely had an opportunity to mate still maintained an increased capacity to survive long-term anoxia relative to age matched hermaprodites indicating that mating and interaction with other males did not compromise the long-term anoxia survival phenotype. The *tra-2(q276)* mutant was used to show that the long-term anoxia survival phenotype observed in males is dependent on male phenotype rather than male genotype. The *tra-2(q276)* mutant is phenotypically male but instead of having the male genotype (X0) is genotypically hermaphroditic (XX). The *tra-2(q276)* animal survived long-term anoxia similar to that of wild-type males indicating that something inherent about the male phenotype confers anoxia tolerance.

Combined, these studies provided further evidence that anoxia tolerance is strongly influenced by physiology and genotype. The ability of an individual to survive anoxic stress is determined by the interplay of multiple pathways in a complex fashion as evidenced by the wide range in biologic function attributed to the many encoded proteins and enzymes recognized to influence anoxia tolerance.

7. The multifactorial architecture of anoxia tolerance

As additional work is conducted to identify the mechanisms by which anoxia tolerant animals survive, recover, and protect tissues it is unlikely that a single important regulatory gene will be identified. Instead, we propose that anoxia survival is by way of a complex interaction of multiple physiological processes that animals are able to survive oxygen-deprivation stress and specifically, anoxia. In this chapter we have discussed a wide range of biological processes that naturally, or in the mutant condition, enhance or reduce anoxia tolerance. We have also related the observation that organisms that survive anoxic stress often have other stress resistant phenotypes as well. For example, the long-term anoxia tolerant strains *glp-1(e2141)* and *daf-2(e1370)* also share an increased longevity phenotype. However, longevity and anoxia-tolerance phenotypes are not superimposable. The extended lifespan of *glp-1* requires the absence of a functional germ line and presence of an intact somatic gonad. In contrast, *daf-2* mutants have full reproductive capacity and have nearly wild-type brood sizes. The long-term anoxia tolerant phenotype is *daf-16*-dependent in *daf-2* mutants, but *daf-16* independent in sterile mutants. The differences in physiology between these two strains are numerous, for example *daf-2* mutants accumulate fat and glycogen while *glp-1* mutants do not, *daf-2* mutants are dauer-constitutive at 25°C but *glp-1* mutants are not. The relationship between sterility-induced anoxia-tolerance and longevity is not yet clear and strains have been identified that carry one but not both characteristics. The unmated *fog-2* mutant has a wild-type lifespan and functional oocytes, yet this mutant is long-term anoxia tolerant unless induced to have offspring by mating. Not all lifespan extended mutant strains have an increase in anoxia tolerance relative to wild-type animals, suggesting that at least an overlap in the mechanisms governing the two phenotypes exists but that they are not identical. Currently, the mechanism by which the germline is regulating anoxia tolerance remains unclear.

Anoxia-tolerance is also under the control of metabolic factors. Reduced signaling through the insulin-IGF pathway confers long-term anoxia tolerance. Animals with reduced caloric intake (which may mimic reduction in insulin signaling) such as the dauer stage of larval development and a mutation in *eat-2* (animals have a reduced pharynx pumping rate and therefore reduced food intake), have been found by our lab to have an elevated anoxia survival rate. Complimentary to this observation is that long-term anoxia tolerant strains have elevated levels of fuel source carbohydrates relative to long-term anoxia sensitive strains. These elevated carbohydrate stores and the associated long-term anoxia tolerance can be suppressed by mutations in *aak-2*, the kinase subunit of the AMPK energy sensor, in some but not all long-term anoxia tolerant strains. The influence of *aak-2* on anoxia tolerance is linked to the activation of cellular stress responses that are under the control of the transcription factor *daf-16*.

The ability to survive extended periods of anoxia is arguably un-adaptive if the animal is unable to resume normal activity such as foraging and reproduction after reoxygenation. It is reasonable to expect that adaptive mechanisms have evolved that protect or repair tissues when damage is incurred during stress. Specific genes have been recognized as required for post-anoxia health and they function in diverse biological processes. For example, *gpd-2* and *gpd-3* are necessary for tissue maintenance during anoxic stress and function in the glycolytic pathway while *hyl-2*, a ceremide synthase required for short-term anoxia survival, functions in a seemingly unrelated manner to provide proper length fatty acyl chains which serve as the precursors of membrane sphingolipids and cell signaling molecules.

The role of environmental factors such as temperature or food source and availability represent yet another genre of factors influencing anoxia tolerance. It is likely that environmental factors exert their influence by altering the rate of reactions associated with the biological processes discussed above. We can view these environmental factors as persistent modern reminders of the pressures to which organisms were obliged to adapt or die. Having evolved under the influence of a range of environmental pressures it is not surprising that multiple mechanisms persist to cope with diverse environmentally induced stresses.

Long-term anoxia survival requires an overall reduction in metabolic rate and ability to provide enough energy to sustain the animal through the oxygen deprivation period and allow maintenance of tissue integrity. Figure 8 depicts a model of the multifactorial character of the anoxia tolerant phenotype. It is likely that a long-term anoxia tolerant strain is able to survive anoxia stress at a high rate due to one or more of the biochemical branches that lead to an anoxia tolerant phenotype. Within each branch specific adaptive responses occur, governed by a particular set of genes that may overlap but are not identical to the set of genes working in the other branches. Therefore, not all long-term anoxia tolerant strains of *C. elegans* survive via a common mechanism.

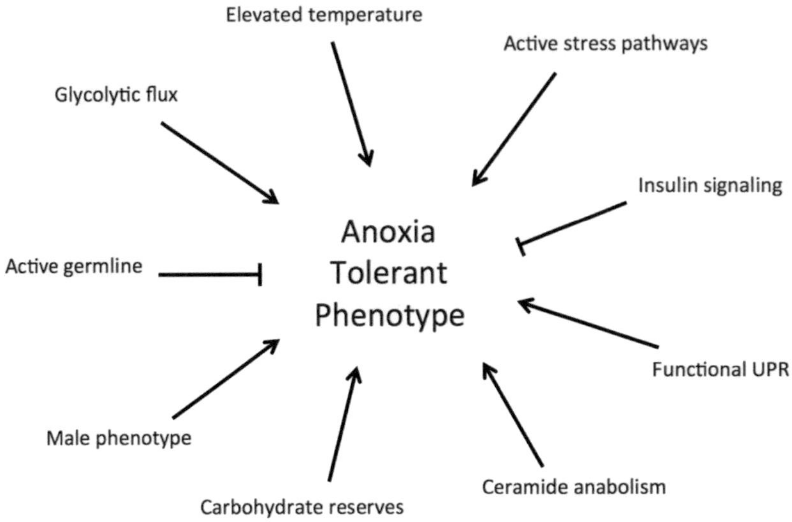

Fig. 8. The anoxia tolerant phenotype is multifactorial in nature. Biological factors that promote enhanced anoxia survival are shown as activating arrows, while factors or conditions that decrease the rate of survival during anoxia exposure are shown as inhibiting blunt-ended lines. We propose that the level of anoxia tolerance for any particular strain is a function of the interaction of the various factors shown in the diagram.

8. Conclusions - suspended animation and human health related issues

Oxygen deprivation is central to many life-threatening human health issues (Semenza, 2010). The extremely high economic and social cost associated with traumas such as blood loss, drowning, suffocation and toxins that affect pulmonary or cardiac function, in addition to diseases that compromise pulmonary and cardiac function underscores the significant importance in understanding responses to oxygen deprivation. In addition to these obvious human health related issues oxygen levels also influence the progression of tumors in individuals afflicted with cancer. It is known that microenvironments within solid tumors exist and that cells in regions of low oxygen are often more resistant to chemotherapeutic or radiation treatment. The solid tumor cells that are further away from the vascular system not only have less chemotherapeutic drugs being delivered but also have a decrease in oxygen levels. This reduction in oxygen can influence the progression of cell division leading to a population of cancer cells that are not rapidly dividing yet remain viable and quiescent. When these cells are re-exposed to oxygen it is possible that they resume rapid cell cycle progression and seed further tumor progression. Therefore, the understanding of how cells, tissues and whole organisms respond to and survive oxygen deprivation is of not only scientific interest but vital in the context of human health related issues.

There are many important and significant approaches that researchers are taking to understand the implications and effects oxygen deprivation has on organisms. Use of *C. elegans* as a genetic model system allows one to characterize many aspects of hypoxia and anoxia responses including the influence on development, cell cycle progression and organ structure and function. Furthermore, the capacity to use cellular and genetic tools to dissect molecular pathways that are involved with oxygen deprivation survival further underscores the value of *C. elegans* as a model system. Finally, the ability to generate a state of anoxia tolerance through genetic manipulation or chemical means will aid in understanding how organisms with complex tissues respond to and survive oxygen deprivation.

The induction of suspended animation in *C. elegans* and zebrafish led to the pursuit of molecules that induce suspended animation in more complex systems (Roth and Nystul, 2005). The general idea is to treat individuals experiencing a traumatic event, such as blood loss leading to severe oxygen deprivation to vital organ systems, by inducing a state of suspended animation so that cellular processes (such as cell death) arrest or slow. Induction of suspended animation may "buy time" until other treatments can be administered. One molecule under intense investigation for inducing a state of suspended animation or a hypometabolic state is hydrogen sulfide (Blackstone et al., 2005; Roth and Nystul, 2005). Remarkably, hydrogen sulfide can reversibly induce a hypometabolic state in which core body temperature can be reduced in mammals (Blackstone et al., 2005; Blackstone and Roth, 2007). Hydrogen sulfide, or molecules with similar capabilities, provides a possible therapeutic approach to treating individuals with life-threatening events that compromise oxygen delivery to vital organs (Aslami et al., 2010; Szabo, 2007; Wagner et al., 2009). Like many new ideas that address biologically and medically complex problems, the ability to use basic sciences from model systems to identify treatments and therapeutics for the benefit of human health related issues is going to be costly, perhaps controversial and quite complex at the biological level (Olson, 2011). However, given the profound effect oxygen deprivation, including anoxia, has on living systems it is of great interest to continue the onward march toward understanding the molecular nature of anoxia tolerance in biological systems.

9. Acknowledgements

We thank the *C. elegans* community for reagents that have been used in various studies that we reviewed here. We thank the *C. elegans* Genetics Stocks Center, the *C. elegans* Knockout Consortium (Oklahoma Medical Research Foundation, the University of British Columbia and The Genome Sciences Center, Vancouver) for strains. Work has been supported by grants form the National Science Foundation and the National Institute of General Medical Sciences (NIH) to PAP. We thank Dr. Gruenbaum for providing Ce-Lamin Antibody and Dr. Landon Moore for the HCP-1 antibody. We gratefully thank Mary Ladage for comments on the chapter and Sarah Goy for assistance with images. Finally, we thank and appreciate input and comments from members of the Padilla Lab and the Developmental Integrative Biology Group at UNT.

10. References

Anderson, G. L. (1978). Responses of dauerlarvae of *Caenorhabditis elegans* (Nematoda: Rhabditidae) to thermal stress and oxygen deprivation. *Can. J. Zool.* 56, 1786-1791.

Anderson, G. L. and Dusenbery, D. B. (1977). Critical-oxygen tension of Caenorhabdiltis elegans. *J Nematol* 9, 253-4.

Apfeld, J., O'Connor, G., McDonagh, T., DiStefano, P. S. and Curtis, R. (2004). The AMP-activated protein kinase AAK-2 links energy levels and insulin-like signals to lifespan in C. elegans. *Genes Dev* 18, 3004-9.

Arantes-Oliveira, N., Apfeld, J., Dillin, A. and Kenyon, C. (2002). Regulation of life-span by germ-line stem cells in Caenorhabditis elegans. *Science* 295, 502-5.
Aslami, H., Heinen, A., Roelofs, J. J., Zuurbier, C. J., Schultz, M. J. and Juffermans, N. P. (2010). Suspended animation inducer hydrogen sulfide is protective in an in vivo model of ventilator-induced lung injury. *Intensive Care Med* 36, 1946-52.
Beale, E. G. (2008). 5'-AMP-activated protein kinase signaling in Caenorhabditis elegans. *Exp Biol Med (Maywood)* 233, 12-20.
Blackstone, E., Morrison, M. and Roth, M. B. (2005). H2S induces a suspended animation-like state in mice. *Science* 308, 518.
Blackstone, E. and Roth, M. B. (2007). Suspended animation-like state protects mice from lethal hypoxia. *Shock* 27, 370-2.
Brenner, S. (1974). The genetics of Caenorhabditis elegans. *Genetics* 77, 71-94.
Brooks, K. K., Liang, B. and Watts, J. L. (2009). The influence of bacterial diet on fat storage in C. elegans. *PLoS One* 4, e7545.
Chalfie, M. and Kain, S. R. (2006). Green Fluorescent Protein Properties, Applications and Protocols: John Wiley & Sons, Inc.
Chen, Q. and Haddad, G. G. (2004). Role of trehalose phosphate synthase and trehalose during hypoxia: from flies to mammals. *J Exp Biol* 207, 3125-9.
Clegg, J. S. (2001). Cryptobiosis--a peculiar state of biological organization. *Comp Biochem Physiol B Biochem Mol Biol* 128, 613-24.
Cremer, T., Kreth, G., Koester, H., Fink, R. H., Heintzmann, R., Cremer, M., Solovei, I., Zink, D. and Cremer, C. (2000). Chromosome territories, interchromatin domain compartment, and nuclear matrix: an integrated view of the functional nuclear architecture. *Crit Rev Eukaryot Gene Expr* 10, 179-212.
Crittenden, S. L., Troemel, E. R., Evans, T. C. and Kimble, J. (1994). GLP-1 is localized to the mitotic region of the C. elegans germ line. *Development* 120, 2901-11.
D'Angelo, M. A., Anderson, D. J., Richard, E. and Hetzer, M. W. (2006). Nuclear pores form de novo from both sides of the nuclear envelope. *Science* 312, 440-3.
D'Angelo, M. A. and Hetzer, M. W. (2008). Structure, dynamics and function of nuclear pore complexes. *Trends Cell Biol* 18, 456-66.
De Souza, C. P., Ellem, K. A. and Gabrielli, B. G. (2000). Centrosomal and cytoplasmic Cdc2/cyclin B1 activation precedes nuclear mitotic events. *Exp Cell Res* 257, 11-21.
Dernburg, A. F. (2001). Here, there, and everywhere: kinetochore function on holocentric chromosomes. *J Cell Biol* 153, F33-8.
DiGregorio, P. J., Ubersax, J. A. and O'Farrell, P. H. (2001). Hypoxia and nitric oxide induce a rapid, reversible cell cycle arrest of the Drosophila syncytial divisions. *J Biol Chem* 276, 1930-1937.
Douglas, R. M., Xu, T. and Haddad, G. G. (2001). Cell cycle progression and cell division are sensitive to hypoxia in Drosophila melanogaster embryos. *Am J Physiol Regul Integr Comp Physiol* 280, R1555-63.
Epstein, A. C., Gleadle, J. M., McNeill, L. A., Hewitson, K. S., O'Rourke, J., Mole, D. R., Mukherji, M., Metzen, E., Wilson, M. I., Dhanda, A. et al. (2001). C. elegans EGL-9 and Mammalian Homologs Define a Family of Dioxygenases that Regulate HIF by Prolyl Hydroxylation. *Cell* 107, 43-54.
Ferrai, C., Jesus de Castro, I., Lavitas, L., Chotalia, M. and Pombo, A. (2011). Gene Positioning. In *The Nucleus,* eds. T. Misteli and D. L. Spector), pp. 115-131. New York: Cold Spring Harbor Laboratory Press.

Foe, V. E. and Alberts, B. M. (1985). Reversible chromosome condensation induced in Drosophila embryos by anoxia: visualization of interphase nuclear organization. *J Cell Biol* 100, 1623-36.

Foll, R. L., Pleyers, A., Lewandovski, G. J., Wermter, C., Hegemann, V. and Paul, R. J. (1999). Anaerobiosis in the nematode Caenorhabditis elegans. *Comp Biochem Physiol B Biochem Mol Biol* 124, 269-80.

Fraser, A. G., Kamath, R. S., Zipperlen, P., Martinez-Campos, M., Sohrmann, M. and Ahringer, J. (2000). Functional genomic analysis of C. elegans chromosome I by systematic RNA interference. *Nature* 408, 325-30.

Frazier, H. N., 3rd and Roth, M. B. (2009). Adaptive sugar provisioning controls survival of C. elegans embryos in adverse environments. *Curr Biol* 19, 859-63.

Galy, V., Askjaer, P., Franz, C., Lopez-Iglesias, C. and Mattaj, I. W. (2006). MEL-28, a novel nuclear-envelope and kinetochore protein essential for zygotic nuclear-envelope assembly in C. elegans. *Curr Biol* 16, 1748-56.

Gems, D. and Riddle, D. L. (2000). Genetic, behavioral and environmental determinants of male longevity in Caenorhabditis elegans. *Genetics* 154, 1597-610.

Gerstbrein, B., Stamatas, G., Kollias, N. and Driscoll, M. (2005). In vivo spectrofluorimetry reveals endogenous biomarkers that report healthspan and dietary restriction in Caenorhabditis elegans. *Aging Cell* 4, 127-37.

Geyer, P. K., Vitalini, M. W. and Wallrath, L. L. (2011). Nuclear organization: taking a position on gene expression. *Curr Opin Cell Biol* 23, 354-9.

Gong, D., Pomerening, J. R., Myers, J. W., Gustavsson, C., Jones, J. T., Hahn, A. T., Meyer, T. and Ferrell, J. E., Jr. (2007). Cyclin A2 regulates nuclear-envelope breakdown and the nuclear accumulation of cyclin B1. *Curr Biol* 17, 85-91.

Gottlieb, S. and Ruvkun, G. (1994). daf-2, daf-16 and daf-23: genetically interacting genes controlling Dauer formation in Caenorhabditis elegans. *Genetics* 137, 107-20.

Gray, J. M., Karow, D. S., Lu, H., Chang, A. J., Chang, J. S., Ellis, R. E., Marletta, M. A. and Bargmann, C. I. (2004). Oxygen sensation and social feeding mediated by a C. elegans guanylate cyclase homologue. *Nature* 430, 317-22.

Greenstein, D. (2005). Control of oocyte meiotic maturation and fertilization: WormBook, ed. The C. elegans Research Community, WormBook, doi/10.1895/wormbook.1.53.1, http://www.wormbook.org.

Hajeri, V. A., Little, B. A., Ladage, M. L. and Padilla, P. A. (2010). NPP-16/Nup50 function and CDK-1 inactivation are associated with anoxia-induced prophase arrest in Caenorhabditis elegans. *Mol Biol Cell* 21, 712-24.

Hajeri, V. A., Trejo, J. and Padilla, P. A. (2005). Characterization of sub-nuclear changes in Caenorhabditis elegans embryos exposed to brief, intermediate and long-term anoxia to analyze anoxia-induced cell cycle arrest. *BMC Cell Biol* 6, 47.

Hardie, D. G., Hawley, S. A. and Scott, J. W. (2006). AMP-activated protein kinase--development of the energy sensor concept. *J Physiol* 574, 7-15.

Hardwick, K. G., Li, R., Mistrot, C., Chen, R. H., Dann, P., Rudner, A. and Murray, A. W. (1999). Lesions in many different spindle components activate the spindle checkpoint in the budding yeast Saccharomyces cerevisiae. *Genetics* 152, 509-18.

Hardwick, K. G. and Murray, A. W. (1995). Mad1p, a phosphoprotein component of the spindle assembly checkpoint in budding yeast. *J Cell Biol* 131, 709-20.

Hartwell, L. H. (2004). Yeast and cancer. *Biosci Rep* 24, 523-44.

Hartwell, L. H. and Weinert, T. A. (1989). Checkpoints: controls that ensure the order of cell cycle events. *Science* 246, 629-34.

Heald, R. and McKeon, F. (1990). Mutations of phosphorylation sites in lamin A that prevent nuclear lamina disassembly in mitosis. *Cell* 61, 579-89.

Herndon, L. A., Schmeissner, P. J., Dudaronek, J. M., Brown, P. A., Listner, K. M., Sakano, Y., Paupard, M. C., Hall, D. H. and Driscoll, M. (2002). Stochastic and genetic factors influence tissue-specific decline in ageing C. elegans. *Nature* 419, 808-14.

Hochachka, P. W. (2000). Oxygen, homeostasis, and metabolic regulation. *Adv Exp Med Biol* 475, 311-35.

Hochachka, P. W., Buck, L. T., Doll, C. J. and Land, S. C. (1996). Unifying theory of hypoxia tolerance: molecular/metabolic defense and rescue mechanisms for surviving oxygen lack. *Proc Natl Acad Sci U S A* 93, 9493-8.

Honda, Y., Tanaka, M. and Honda, S. (2010). Trehalose extends longevity in the nematode Caenorhabditis elegans. *Aging Cell.*

Hottiger, T., De Virgilio, C., Hall, M. N., Boller, T. and Wiemken, A. (1994). The role of trehalose synthesis for the acquisition of thermotolerance in yeast. II. Physiological concentrations of trehalose increase the thermal stability of proteins in vitro. *Eur J Biochem* 219, 187-93.

Hsin, H. and Kenyon, C. (1999). Signals from the reproductive system regulate the lifespan of C. elegans. *Nature* 399, 362-6.

Jiang, H., Guo, R. and Powell-Coffman, J. A. (2001). The Caenorhabditis elegans hif-1 gene encodes a bHLH-PAS protein that is required for adaptation to hypoxia. *Proc Natl Acad Sci U S A* 98, 7916-21.

Jorgensen, E. M. and Mango, S. E. (2002). The art and design of genetic screens: caenorhabditis elegans. *Nat Rev Genet* 3, 356-69.

Kenyon, C., Chang, J., Gensch, E., Rudner, A. and Tabtiang, R. (1993). A C. elegans mutant that lives twice as long as wild type. *Nature* 366, 461-4.

Kenyon, C. J. (2010). The genetics of ageing. *Nature* 464, 504-12.

Kimura, K. D., Tissenbaum, H. A., Liu, Y. and Ruvkun, G. (1997). daf-2, an insulin receptor-like gene that regulates longevity and diapause in Caenorhabditis elegans. *Science* 277, 942-6.

Larsen, P. L. (2001). Asking the age-old questions. *Nat Genet* 28, 102-4.

Larsen, P. L., Albert, P. S. and Riddle, D. L. (1995). Genes that regulate both development and longevity in Caenorhabditis elegans. *Genetics* 139, 1567-83.

LaRue, B. L. and Padilla, P. A. (2011). Environmental and genetic preconditioning for long-term anoxia responses requires AMPK in Caenorhabditis elegans. *PLoS One* 6, e16790.

Lee, D. L. (1965). The Physiology of Nematodes. San Francisco: W.H. Freeman and Company.

Lee, K. K., Gruenbaum, Y., Spann, P., Liu, J. and Wilson, K. L. (2000). C. elegans nuclear envelope proteins emerin, MAN1, lamin, and nucleoporins reveal unique timing of nuclear envelope breakdown during mitosis. *Mol Biol Cell* 11, 3089-99.

Lin, K., Dorman, J. B., Rodan, A. and Kenyon, C. (1997). daf-16: An HNF-3/forkhead family member that can function to double the life-span of Caenorhabditis elegans. *Science* 278, 1319-22.

Lindqvist, A., Rodriguez-Bravo, V. and Medema, R. H. (2009). The decision to enter mitosis: feedback and redundancy in the mitotic entry network. *J Cell Biol* 185, 193-202.

Lindqvist, A., van Zon, W., Karlsson Rosenthal, C. and Wolthuis, R. M. (2007). Cyclin B1-Cdk1 activation continues after centrosome separation to control mitotic progression. *PLoS Biol* 5, e123.

McElwee, J., Bubb, K. and Thomas, J. H. (2003). Transcriptional outputs of the Caenorhabditis elegans forkhead protein DAF-16. *Aging Cell* 2, 111-21.

McGhee, J. D. (2007). The C. elegans intestine. In *WormBook*, (ed. M. Chalfie): The C. elegans Research Community.

Mehrani, H. and Storey, K. B. (1995). Enzymatic control of glycogenolysis during anoxic submergence in the freshwater turtle Trachemys scripta. *Int J Biochem Cell Biol* 27, 821-30.

Mendelsohn, B. A., Kassebaum, B. L. and Gitlin, J. D. (2008). The zebrafish embryo as a dynamic model of anoxia tolerance. *Dev Dyn* 237, 1780-8.

Mendenhall, A. R., LaRue, B. and Padilla, P. A. (2006). Glyceraldehyde-3-phosphate dehydrogenase mediates anoxia response and survival in Caenorhabditis elegans. *Genetics*.

Mendenhall, A. R., LeBlanc, M. G., Mohan, D. P. and Padilla, P. A. (2009). Reduction in ovulation or male sex phenotype increases long-term anoxia survival in a daf-16-independent manner in Caenorhabditis elegans. *Physiol Genomics* 36, 167-78.

Menuz, V., Howell, K. S., Gentina, S., Epstein, S., Riezman, I., Fornallaz-Mulhauser, M., Hengartner, M. O., Gomez, M., Riezman, H. and Martinou, J. C. (2009). Protection of C. elegans from anoxia by HYL-2 ceramide synthase. *Science* 324, 381-4.

Miller, D. L. and Roth, M. B. (2009). C. elegans are protected from lethal hypoxia by an embryonic diapause. *Curr Biol* 19, 1233-7.

Miller, M. A., Nguyen, V. Q., Lee, M. H., Kosinski, M., Schedl, T., Caprioli, R. M. and Greenstein, D. (2001). A sperm cytoskeletal protein that signals oocyte meiotic maturation and ovulation. *Science* 291, 2144-7.

Moore, L. L., Morrison, M. and Roth, M. B. (1999). HCP-1, a protein involved in chromosome segregation, is localized to the centromere of mitotic chromosomes in Caenorhabditis elegans. *J Cell Biol* 147, 471-80.

Murphy, C. T., McCarroll, S. A., Bargmann, C. I., Fraser, A., Kamath, R. S., Ahringer, J., Li, H. and Kenyon, C. (2003). Genes that act downstream of DAF-16 to influence the lifespan of Caenorhabditis elegans. *Nature* 424, 277-83.

Nurse, P., Masui, Y. and Hartwell, L. (1998). Understanding the cell cycle. *Nat Med* 4, 1103-6.

Nystul, T. G., Goldmark, J. P., Padilla, P. A. and Roth, M. B. (2003). Suspended animation in C. elegans requires the spindle checkpoint. *Science* 302, 1038-41.

Nystul, T. G. and Roth, M. B. (2004). Carbon monoxide-induced suspended animation protects against hypoxic damage in Caenorhabditis elegans. *Proc Natl Acad Sci U S A* 101, 9133-6.

O'Farrell, P. H. (2001). Conserved responses to oxygen deprivation. *J Clin Invest* 107, 671-4.

Oegema, K., Desai, A., Rybina, S., Kirkham, M. and Hyman, A. A. (2001). Functional analysis of kinetochore assembly in Caenorhabditis elegans. *J Cell Biol* 153, 1209-26.

Oliveira, R. P., Porter Abate, J., Dilks, K., Landis, J., Ashraf, J., Murphy, C. T. and Blackwell, T. K. (2009). Condition-adapted stress and longevity gene regulation by Caenorhabditis elegans SKN-1/Nrf. *Aging Cell* 8, 524-41.

Olson, K. R. (2011). The Therapeutic Potential of Hydrogen Sulfide: Separating Hype from Hope. *Am J Physiol Regul Integr Comp Physiol*.

Padilla, P. A., Nystul, T. G., Zager, R. A., Johnson, A. C. and Roth, M. B. (2002). Dephosphorylation of Cell Cycle-regulated Proteins Correlates with Anoxia-induced Suspended Animation in Caenorhabditis elegans. *Mol Biol Cell* 13, 1473-83.

Padilla, P. A. and Roth, M. B. (2001). Oxygen deprivation causes suspended animation in the zebrafish embryo. *Proc Natl Acad Sci U S A* 12, 12.

Paul, R. J., Gohla, J., Foll, R. and Schneckenburger, H. (2000). Metabolic adaptations to environmental changes in Caenorhabditis elegans. *Comp Biochem Physiol B Biochem Mol Biol* 127, 469-79.

Podrabsky, J. E., Lopez, J. P., Fan, T. W., Higashi, R. and Somero, G. N. (2007). Extreme anoxia tolerance in embryos of the annual killifish Austrofundulus limnaeus: insights from a metabolomics analysis. *J Exp Biol* 210, 2253-66.

Powell-Coffman, J. A. (2010). Hypoxia signaling and resistance in C. elegans. *Trends Endocrinol Metab* 21, 435-40.

Renfree, M. B. and Shaw, G. (2000). Diapause. *Annu Rev Physiol* 62, 353-75.

Riddle, D. L. (1988). The Dauer Larva. In *The nematode Caenorhabditis elegans*, (ed. W. Wood), pp. 393-412. Plainview: Cold Spring Harbor Laboratory Press.

Riddle, D. L., Swanson, M. M. and Albert, P. S. (1981). Interacting genes in nematode dauer larva formation. *Nature* 290, 668-71.

Roth, M. B. and Nystul, T. (2005). Buying time in suspended animation. *Sci Am* 292, 48-55.

Scott, B. A., Avidan, M. S. and Crowder, C. M. (2002). Regulation of hypoxic death in C. elegans by the insulin/IGF receptor homolog DAF-2. *Science* 296, 2388-91.

Semenza, G. L. (2007). Life with oxygen. *Science* 318, 62-4.

Semenza, G. L. (2010). Oxygen homeostasis. *Wiley Interdiscip Rev Syst Biol Med* 2, 336-61.

Singer, M. A. and Lindquist, S. (1998). Multiple effects of trehalose on protein folding in vitro and in vivo. *Mol Cell* 1, 639-48.

Smitherman, M., Lee, K., Swanger, J., Kapur, R. and Clurman, B. E. (2000). Characterization and targeted disruption of murine Nup50, a p27(Kip1)-interacting component of the nuclear pore complex. *Mol Cell Biol* 20, 5631-42.

Suda, H., Shouyama, T., Yasuda, K. and Ishii, N. (2005). Direct measurement of oxygen consumption rate on the nematode Caenorhabditis elegans by using an optical technique. *Biochem Biophys Res Commun* 330, 839-43.

Szabo, C. (2007). Hydrogen sulphide and its therapeutic potential. *Nat Rev Drug Discov* 6, 917-35.

Timmons, L. and Fire, A. (1998). Specific interference by ingested dsRNA. *Nature* 395, 854.

Tissenbaum, H. A. and Ruvkun, G. (1998). An insulin-like signaling pathway affects both longevity and reproduction in Caenorhabditis elegans. *Genetics* 148, 703-17.

Van Voorhies, W. A. and Ward, S. (2000). Broad oxygen tolerance in the nematode *Caenorhabditis elegans*. *J Exp Biol* 203 Pt 16, 2467-78.

Vanfleteren, J. R. and De Vreese, A. (1996). Rate of aerobic metabolism and superoxide production rate potential in the nematode Caenorhabditis elegans. *J Exp Zool* 274, 93-100.

Wagner, F., Asfar, P., Calzia, E., Radermacher, P. and Szabo, C. (2009). Bench-to-bedside review: Hydrogen sulfide--the third gaseous transmitter: applications for critical care. *Crit Care* 13, 213.

WormBook. WormBook, eds. M. Chalfie and J. Mendel)

Part 2

Human Health Related Issues

3

Resistance of Multipotent Mesenchymal Stromal Cells to Anoxia *In Vitro*

L.B. Buravkova, E.R. Andreeva, J.V. Rylova and A.I. Grigoriev
Institute of Biomedical Problems, Russian Academy of Sciences,
Faculty of Basic Medicine MSU, Moscow,
Russia

1. Introduction

Oxygen balance is a corner element of tissue physiology. The damaging effects of oxygen deprivation have been under consideration over at least 100 years. Recently, interest to hypoxia and practically complete absence of oxygen referred to as anoxia has been rekindled in context of the great progress made in the studies of stem/progenitor cell function in organism. The hallmark of stem cells is ability to self-renew and maintain multipotency. This ability depends on the balance of complex signals in their microenvironment. One of the most important findings is that oxygen represents a crucial component determining stem cells homeostasis within their native tissue niche. The stem cell niche has come to refer to a defined anatomical compartment that includes cellular and acellular components that integrate both systemic and local cues to regulate the stem cells biology [Jones and Wagers, 2008; Li and Xie, 2005; Scadden, 2006; Yin and Li, 2006; Buravkova & Andreeva, 2010]. The first specialized tissue niche was described for hematopoietic cells in bone marrow [Schofield, 1978]. Cells, blood vessels, matrix glycoproteins, and the three-dimensional space formed the architecture of a highly specialized microenvironment for stem cells [Scadden, 2006]. Oxygen measurements in tissues known to harbor stem cells revealed low level of oxygen, and raised the question of whether such an environment was necessary for the niche to maintain stem cells [Braun et al., 2001; Cipolleschi et al., 1993; Erecinska and Silver, 2001]. Recent evidence has broadened the spectrum of stem cells influenced by limited oxygen supply including cancer stem cells and induced pluripotent stem cells [Brahimi-Horn & Pouysse'gur, 2007]. Low oxygen tension maintains the undifferentiated state of embryonic, hematopoietic, mesenchymal, and neural stem cell phenotypes, and affects proliferation and cell-fate commitment [Mohyeldin et al., 2010].

Multipotent mesenchymal stem/stromal cells (MMSCs) arouse interest of cell biologists because of high proliferating activity and multilineage differentiion capacity. These cells are also shown to be immunoprivilege and to possess immunosuppressive features. The MMSC properties taken together make these cells a very attractive tool for cell therapy and regenerative medicine. By and large, manifestation of the MMSC properties is strictly dependent on oxygen concentration in native milieu. Moreover, the best realization of the regenerative potential is closely associated with low or very low oxygen level in the area of tissue damage.

The chapter highlights the recent progress in evaluation of the pivotal role of low oxygen in MMSC milieu and how it uniquely modulates the MMSC properties.

2. MMSCs and microenvironmental requirements: the role of oxygen

MMSCs (a rare population of non-hematopoietic stem/progenitor cells) are the subject of increasing scientific interest due to the key role they play in physiological renewal and repair. For a long time there was only one special tissue known as a definite source for renewal and substitution of cells in mammals, humans in particular. It was bone marrow capable to produce new mature blood cells from undifferentiated hematopoietic precursors. Marrow stroma contains many cell elements including endothelial cells of vessels, reticular cells, fibroblasts, adipocytes, stromal cells, and macrophages. Among the multiple stromal cells there is a minor population of MMSCs that localizes assumingly in perivascular regions of the bone marrow [Fridenshtein et al., 1976]. According to the modern concept, this population has the capacity to differentiate into cells of mesenchymal lineage (osteoblasts, chondroblasts, adipocytes and some other types of cells) [Kolf et al., 2007; Losito et al., 2009; Mohyeldin et al., 2010].

The direct evidence of stromal progenitor cells entity *in vivo* was not available due to the lack of a single definitive marker; therefore, demonstration of their existence has relied primarily on retrospective *in vitro* assays. To date, identification and characterization of bone marrow MMSCs from various animal species and humans have been described in numerous papers. It was recognized that the main phenotypic MMSC features should satisfy the following three basic criteria: adhesion to plastic, extended self-maintenance in culture, and the capacity to differentiate into bone, cartilage, adipose and hematopoiesis-inducing stroma during transplantation *in vivo* or upon certain inductive stimuli *in vitro* [Caplan, 2007; Kolf et al., 2007].

As it has been already mentioned, firstly MMSCs were described in bone marrow. Subsequently cells with characteristics similar to MMSCs were isolated from other tissue sources, including trabecular bone, adipose tissue, synovium, skeletal muscle, lung, deciduous teeth, and human umbilical cord perivascular cells derived from the Wharton's Jelly, peripheral blood, dental pulp, periodontal ligament and etc. (Tabl. 1) [for refs. see also Kolf et al., 2007; Augello et al., 2010].

These findings reveal that MMSCs are diversely distributed *in vivo* and, as a result, may occupy a ubiquitous stem cell niche. There is a hypothesis that these cells are the common source of multipotent cells in adult organism migrating constantly in various mesenchymal tissues and providing their maintenance, renewal and regeneration.

The stem/progenitor cell microenvironment constists of specific molecular, cellular, and physiological components and is subject to physical and mechanical stimuli. Although stem cells can reside in markedly different locations and have distinctly different developmental paths, low oxygen tension (termed hypoxia or, in case of extremely low O_2, anoxia) seems to be a common *in vivo* feature shared by many types of adult stem cells. Indeed, there is increasing evidence that presence/absence of oxygen is a powerful tool that regulates stem cell proliferation and differentiation [Ma et al., 2009].

Tissue source	Representative References
Bone marrow	Bruder et al., 1998; Pittenger et al., 1999; Makino et al., 1999; Majumdar et al., 2000; Bianco et al., 2001; Shake et al., 2002; Shi & Grontos, 2003; Lee et al., 2004; Wagner et al., 2005; Romanov et al., 2006; Fehrer et al., 2007
Adipose tissue	Zuk et al., 2001, 2002; Lee et al., 2004; Rehman et al., 2004; Wagner et al., 2005; Romanov et al., 2006; Gimble et al., 2007; Schaffler et al., 2007; Buravkova et al., 2009; Madonna et al., 2009
Muscle	Bosch et al., 2000; Black, 2001
Umbilical cord blood	Wagner et al., 2005; Caballero et al., 2010
Peripheral blood	Zvaifler et al., 2000
Dermis	Black, 2001
Periosteum	Caballero et al., 2010
Dental pulp	Shi & Grontos, 2003
Synovial membrane	Kurose et al., 2010

Table 1. MMSC tissue sources in humans.

The role of oxygen in maintaining of both self-renewal and committed status of hematopoietic cells has been described in detail [Ivanovic, 2009; Eliasson & Jonasson, 2010, Valtieri & Sorrenino, 2008]. In bone marrow, the hematopoietic compartments are bound by stromal elements [Kolf et al., 2007], mainly MMSCs, and such way that two cell types form an integral part of each other's niche. Although the importance of understanding the progenitor cell spatial distribution with respect to oxygen supply from blood vessel has long been recognized, a direct noninvasive *in vivo* measurement of spatial oxygen gradient in bone marrow has been a major technical hurdle. Early direct measurements revealed that bone marrow is generally hypoxic with O_2 in some regions as low as ~1-2% [Cipolleschi et al., 1993] and even close to anoxia 0.1% O_2 in the osteoblastic niche [Calvi et al., 2003]. Results from the recent *in vivo* studies provided a direct experimental evidence that long-term repopulating HSCs in mouse reside in hypoxic environment [Parmar et al., 2007] and that hypoxia may in fact be an essential part of microenvironment maintaining cells in the undifferentiated state. On the other hand, hypoxia increases erythropoiesis - one of hematopoietic lineage, - by EPO or, maybe, low O_2 [Vlaski et al., 2009].

Based on these data, one may assume that MMSC physiology as an integral part of HSCs niche is also governed mainly by hypoxic and even anoxic conditions. Unlike HSCs residing exactly in bone marrow, MMSCs localize in other perivascular tissue depots and, being involved in regenerative and reparative processes are faced with different oxygen conditions; therefore, they should possess a high degree of O_2-mediated plasticity. Low/extremely low (hypoxia/anoxia) oxygen partial pressure in extracellular space may be physiologic or damaging in consequence of insufficient blood supply to impaired tissue. Low oxygen can modify drastically morphologic and functional cell properties, such as viability, proliferative status, immunophenotype, and differentiation. On the other hand, it is known that different cells have different tolerance to low oxygen [Csete, 2005; Ivanovic, 2009]. MMSCs, a mixture of stem/progenitor cells that may come to be in different oxygen milieu *in vivo,* are an attractive experimental model to explore the intrinsic mechanisms of cell adaptation to oxygen limitation.

The major bulk of knowledge concerning mesenchymal stem/progenitors biology came from the *in vitro* studies. The high proliferative activity underlies the MMSC ability to self-renew in culture for an extended period without a dramatic decline of the telomerase activity and change of karyotype [Bruder et al., 1997; Izadpanah et al., 2006], and to form an uniform layer of adhesive spindle-shaped cells with typical fibroblast-like morphology *in vitro* in the normalized conditions, i.e. low glucose content, absence of differentiation stimuli and appropriate seeding density [Pittenger et al., 1999]. The composition of the gaseous phase in cell culture technique is the most conservative parameter and exploring atmospheric O_2 concentration. It should be taken into account that oxygen concentration is significantly lower *in vivo*. Arterial blood contains about 12% oxygen, and the mean tissue level of oxygen is about 3% with considerable local and regional variation. These are values for adult organs and tissues. Mean oxygen tension in embryonic tissue (where stem cells are enriched relative to adult tissues) is considerably less. Although many papers refer to low oxygen levels in embryos as "hypoxic," they are actually "normoxic" for the time and place of development [Csete, 2005]. In the last decade it was demonstrated that low oxygen tension influences greatly biology of both embryonic and adult stem cell *in vitro* [Eliasson and Jonsson, 2010; Panchision, 2009; Silvan et al., 2009] improving the proliferative and migrating abilities and reducing differentiation and proapoptotic reaction of stem cells. These observations fueled a hypothesis that low oxygen tension could be critical, but not damaging to stem cells microenvironment.

Nowadays, *in vitro* studies of the oxygen effect on MMSC functional properties are growing in number. The initial theory that the replicating stem cell microenvironment should provide sufficient oxygen supply to support tissue growth evolved to the understanding of a more complex signaling role of oxygen in regulation of stem cells migration, differentiation, and development [Ma et al., 2009]. Despite the claim that low and extremely low O_2 may be more representative of the physiological conditions for certain cell types than so-called "normoxia", low oxygen is traditionally called hypoxia in consistency with the conventional terminology.

At the moment, there is wealth of data concerning low oxygen effects on the functional properties of MMSCs *in vitro*. It was demonstrated clearly that hypoxia of different severity induces proliferation in cultured MMSCs from various species. Thus, an increased proliferation rate was demonstrated for rat bone marrow MMSCs at 5% O_2 [Lennon et al., 2001; Buravkova & Anokhina, 2007,2008]. Also, proliferation of human bone marrow MMSCs was stimulated by 2% O_2 [Grayson et al., 2006], 3% 0_2 [Fehrer et al., 2007; D'Ippolito et al., 2006], and 5% O_2 [Zhambalova et al., 2010]. Accelerated cell growth was observed in pig bone marrow MMSC culture at 5% O_2 [Bosch et al, 2006] and murine's bone marrow MMSCs at 8% O_2 [Ren et al., 2006]. Villarruel S.M. and coauthors [2008] estimated the human bone marrow MMSC colony-forming potential at 1, 5, 10, and 20 % O_2, and found that the number of CFU-F raised most at 5% O_2. The data on MMSCs under low oxygen pressure are also discussed in detail in several review papers [Malda et al., 2007; Ma et al., 2009; Das et al., 2010].

Under reduced oxygen pressure MMSCs also display angiogenic activity. At 1% O_2, murine's bone marrow MMSCs migrated rapidly, formed a three-dimensional capillary-like structure in Matrigel, and synthesized more vascular endothelial growth factor (VEGF); matrix metalloproteinase (MMP)-2 mRNA expression and protein secretion were down regulated, while those of membrane-type (MT)MMP-1 were strongly induced by hypoxia [Annabi et al., 2003]. The capillary-like structures were also demonstrated in hypoxic (5%) cultures of human marrow MMSCs [Zhambalova et al., 2010] and adipose tissue MMSCs [Grinakovskaya, personal communication].

MMSCs are considered as a perspective tool for regenerative medicine and approaches to improve MMSCs quality are being developed rapidly. The preconditioning in low oxygen medium is one of the attractive ways. It was demonstrated that preconditioning of human marrow-derived MMSCs in 1%-3% oxygen activated the Akt-signaling pathway while maintaining cell viability and cell cycle rates, induced expression of cMet, the major receptor for hepatocyte growth factor (HGF), and enhanced cMet signaling. MMSCs cultured in hypoxic conditions increased migration rate. Preconditioned normoxic and hypoxic MMSCs equally improved revascularization after surgical hind limb ischemia; however, restoration of blood flow was observed significantly earlier in mice that had been injected with hypoxic preconditioned MMSCs [Rosova et al., 2008]. According to Hu et al. [2008], subletally hypoxic close to anoxia (0.5%) preconditioning of murine bone marrow MMSCs increased expression of prosurvival and proangiogenic factors including hypoxia-inducible factor 1, angiopoietin-1, vascular endothelial growth factor and its receptor, Flk-1, erythropoietin, Bcl-2, and Bcl-xL. Caspase-3 activation in hypoxic MMSCs and populaltion of apoptotic cells were significantly lower compared with normoxic cells *in vitro*. Transplantation of hypoxic vs normoxic MMSCs after myocardial infarction resulted in an increase in angiogenesis, as well as enhanced morphologic and functional benefits of stem cell therapy [Hu et al., 2008].

The adipose tissue-derived MMSCs under low oxygen pressure are of special interest because of the considerable promise for regenerative medicine and cell therapy. Most of the data on MMSCs at low O_2 were gathered using bone marrow MMSCs. Much less investigations have been concerned with the hypoxia effects on stromal cells derived from adipose tissue.

In a few papers hypoxia has been shown to affect the differentiation potential of the adipose tissue-derived MMCSs. Wang et al. [2005] demonstrated that human adipose MMSCs in alginate beads did not display proliferative activity at 5% O_2 in normal expansion medium; however in chondrogenic medium its growth rates was lower at 5% in comparison with 20% O_2. Still, under these conditions they exhibited enhanced chondrogenic differentiation markers including collagen II, glucosaminoglycan, and chondroitin-4-sulfate production [Wang et al., 2005]. The hypoxia effect on adipose tissue MMSCs is strongly dependent on the cultivation conditions. For this reason, there are conflicting data regarding chondrogenic gene expression during induction under hypoxic conditions [Khan et al., 2007; Betre et al., 2006]. Adipose tissue MMSCs expanded in 20% O_2 and transferred into a 2% O_2 environment failed to differentiate robustly to either adipogenic or osteogenic lineages as compared with adipose MMSCs differentiated in normal atmospheric conditions [Lee & Kemp, 2006].

We have developed an experimental approach utilizing permanent expansion of adipose tissue derived MMSCs at a reduced oxygen tension [Buravkova et al., 2009]. In hypoxia (5% O_2) MMSCs demonstrated enhanced growth exceeding that in normoxia (20% O_2) in 2.9 ± 0.2 folds ($p<0.05$) [Buravkova et al., 2009]. The osteogenic differentiation capacity of MMSCs was significantly reduced in hypoxia vs normoxia [Grinakovskaya et al., 2009].

After expansion at low oxygen (2%) adipose tissue derived MMSCs were able to enhance the wound-healing function. Conditioning medium of hypoxic MMSCs promoted significantly collagen synthesis and migration of human dermal fibroblasts in vitro, and reduced the wound area in animal studies. These effects were based on up-regulation of growth factors such as the vascular endothelial growth factor (VEGF) and basic fibroblast growth factor (bFGF) [Lee et al., 2009].

The data above demonstrated clearly that low oxygen concentration in MMSCs microenvironment *in vitro* leads to modulation of MMSCs functions rather than impairment. The question arises whether extremely low oxygen (~ 0% O_2) brings damage to MMSCs?

Papers dedicated to MMSCs and anoxia are few and far between. Nevertheless, the data on the "true" anoxia effects on MMSCs *in vitro* do exist. For example, studies of Annexin V-positive cells in rat bone marrow MMSC culture during 12 h anoxia exposure revealed a time-dependent increase in apoptotic cells from 3% to 15%. When following a 3 h anoxic exposure these MMSCs (apoptotic rate ~ 4%) were cocultured with rat's cardiomyocytes or injected into infarcted zone in the heart, cardiomyocyte death reduced significantly owing to the treatmeant with both normoxic MMSCs and anoxic MMSCs, the Bcl-2/Bax protein ratio increased and cleaved cysteine-aspartic acid protease-3 decreased; anoxic MMSCs were superior to MMSCs in the normoxic condition. Consequently, MMSCs exert the antiapoptotic effect on cardiomyocytes, partially by paracrine action. The authors assume that anoxic preconditioning may be an effective and convenient way to enhance the cardioprotective effect of MMSCs [He et al., 2009].

3. Resistance of rat's marrow MMSCs to extremely low oxygen

Cell morphology and immunophenotype. We have examined the direct effects of 96 h anoxia (~0%O_2) on rat bone marrow MMSCs. Anoxia did not affect cell morphology. The percentage of MMSCs bearing CD90, CD54, CD44, CD29 (more than 95%), CD45, CD11b (less than 0.6%) molecules, and pattern of molecules expression were identical in normoxic and anoxic cells. The slight reduction in percentage of positive cells at anoxia was observed only for the CD73 marker (72% in normoxia vs 67% in anoxia).

Cell growth. Assessment of bone marrow MMSC proliferation in 1-4th passages did not reveal inhibition of the MMSC proliferative activity [Fig. 1; Anokhina et al., 2009]. It should be mentioned that during MMSC exposure in hypoxia proliferative rate displayed a more pronounced excess over normoxic MMSCs [Buravkova et al., 2007] as compared to anoxic MMSCs. Nevertheless, the fact that MMSCs can really proliferate in anoxia seems amazing and deserves special attention.

Absence of evidence for cell proliferation inhibition/deceleration in anoxia disagrees with the results of experiments demonstrating the cell cycle arrest in the conditions of anoxia [Amellem et al., 1994; Gardner et al., 2001; Goda et al., 2003].

The mechanism of inactivation of enzymes involved in nucleic acids production and subsequent inhibition of DNA replication was proposed as an explanation of the phenomenon of the cell growth arrest in murine embryonic fibroblasts and splenic B lymphocytes in low oxygen environment. Moreover, this held true for total oxygen deprivation (0.01%) or anoxia only, but not for 0.1-1% O_2 [Goda et al., 2003]. The drop in bromodeoxyuridine incorporation into murine embryonic fibroblast DNA also confirmed growth arrest due to blockade of replication in anoxia [Gardner et al., 2001]. At very low O_2 (0.01-0.13%) NHIK 3025 cells were able to enter into S-phase of cell cycle but failed to complete DNA synthesis [Amellem et al., 1994].

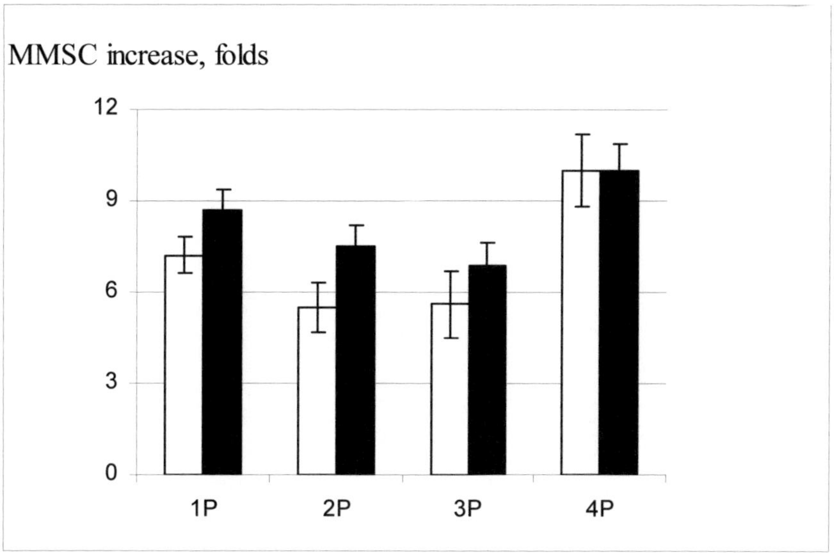

Fig. 1. Rat bone marrow MMSC growth after 96 h of exposure in 20% O_2 - □, and ~0% O_2 - ■ on 1st-4th passages. The averaged data of 7 independent experiments are presented as M+m.

However, in anoxic conditions MMSCs showed a normal process of cell division. It is known that different cell types are characterized by varying resistance to O_2 deprivation in microenvironment [Ivanovic, 2009; Mohyeldin et al., 2010]. If a trace amount of O_2 remaining in pericellular space after gas force-out is sufficient for DNA replication, cell division can be completed successfully, otherwise cells will undergo apoptosis. On the other hand, it was shown that fibroblasts are able to recover after the cell cycle arrest in anoxic conditions [Gardner et al., 2001]. To sum up, cellular mechanisms underlying MMSCs proliferation in anoxia are not clear.

Cell viability. The data on MMSCs viability in normoxic vs anoxic conditions are presented on Fig. 2. Though the share of damaged (apoptotic+necrotic) cells in anoxia did not differ from that in normoxic conditions significantly, the rate of apoptotic cells was doubled in anoxia ($p<0.05$) (Fig. 2).

Therefore, 96 h anoxia did not lead to a considerable increase in the number of damaged cells. At the same time, anoxia appeared to induce apoptosis in MMSCs. It is believed that apoptosis may be associated with anoxia, since a significant O_2 reduction increases frequency of point mutations accumulation of which can be prevented by cell death through the apoptotic path [Greijer & van der Wall, 2004]. Apoptosis due to O_2 deprivation can be triggered by different mechanisms. The key role is played by ROS, JNK kinase, and cytochrome C release from mitochondria mediated, in its turn, by various factors including p53. p53 induction may result from stabilization of HIF-1α protein, the key transcription factor involved in hypoxia and capable to start other, p53-indepent mechanisms of apoptosis [Greijer & van der Wall, 2004]. Besides apoptosis, O_2 deprivation can mobilize some other cell death pathways based on indirect effects of O_2 reduction. Intracellular acidosis triggered by O_2 deprivation is the most important one [Schmaltz et al., 1998].

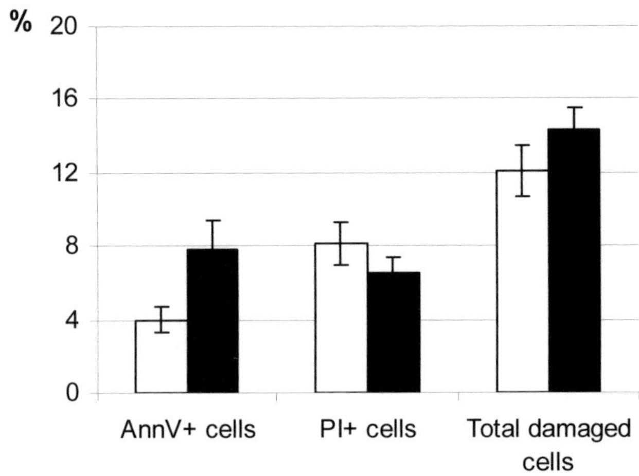

Fig. 2. Rat bone marrow MMSC viability after 96 h of exposure in 20% O_2 - □, and ~0% O_2 - ■. AnnV+ cells – MMSCs, stained with Annexin V-FITC. PI+ cells – MMSCs stained with Propidium Iodide. Total damaged cells – (AnnV+ cells) + (PI+ cells). The averaged data of 12 independent experiments are presented as M±m.

Earlier we described the antiapoptotic effects of hypoxia on rat bone marrow MMSCs [Buravkova et al., 2007]. The antiapoptotic effect of hypoxia and proapoptotic effect of anoxia may be explained using the data on the impact of HIF phosphorylation level on cell viability. The dephosphorylated HIF-1α subunit may indicate the proapoptotic HIF effect through p53 binding, whereas phosphorylated HIF-1α does not [Suzukiet al., 2001]. It may also come in line with the data suggesting that the proapoptotic effect of low oxygen depends on HIF stabilization occurring mostly at less than 5% oxygen, while the antiapoptotic hypoxia effect takes place regardless of HIF [Jiang, Semenza, Bauer 1996; Greijer & van der Wall, 2004].

Thus, path of cell death may be determined by severity of O_2 deprivation which is illustrated by the opposing apoptotic trends in hypoxia and anoxia.

Despite the relative MMSC resistance to 96 h anoxia, a prolonged exposure (up to 3 weeks) to anoxic conditions resulted in a significant reduction in cells viability (Tabl. 2). During 3 weeks in culture the percent of damaged MMSCs in normoxia and hypoxia varied insignificantly (Tabl. 2). Prolonged MMSCs exposure in anoxia led to a drastic decrease in viability up to 22% after one week, 50% after 2 weeks and to practically total cell death after

3 weeks. Necrosis was the predominant path of cell death in anoxic MMSCs, though the percentage of apoptotic cells increased also significantly (Tab.2).

After 1st week	*AnnV+ (%)*	*PI + (%)*	*Damaged cells, total (%)*
Normoxia (20% O_2)	5,7 ± 0,7	5,0 ± 0,2	10,8 ± 0,9
Hypoxia (5% O_2)	4,9 ± 0,8	10,0 ± 0,7	14,9 ± 0,1
Anoxia (~0% O_2)	8,1 ± 0,7	13,9 ± 0,6	22,0
After 2 weeks			
Normoxia (20% O_2)	1,2 ± 0,2	4,9 ± 0,04	6,2 ± 0,2
Hypoxia (5% O_2)	3,0 ± 0,01	10,3 ± 0,5	13,3 ± 0,5
Anoxia (~0% O_2)	16,8 ± 0,4	42,2 ± 2,0	58,9 ± 1,6
After 3 weeks			
Normoxia (20% O_2)	1,4 ± 0,4	8,2 ± 0,1	9,7 ± 0,5
Hypoxia (5% O_2)	0,4 ± 0,1	7,4 ± 0,4	7,8 ± 0,3
Anoxia (~0% O_2)	8,2 ± 0,4	89,0 ± 0,7	97,2 ± 1,1

The averaged data of 5 independent experiments are presented as M±m.

Table 2. Rat bone marrow MMSC viability after prolonged exposure in normoxia, hypoxia and anoxia.

The predominance of necrosis over apoptosis in MMSCs in the course of prolonged exposure to anoxia seems quite natural. In contrast to necrosis, apoptosis is a programmed energy-dependent cell death and, therefore, cell death path is determined by energy state of cells [Leist et al., 1997].

Consequently, MMSC death due to short-term anoxia is minimal and realized primary through the apoptotic pathway. Prolonged anoxic exposure induced massive cell death associated mainly with necrosis.

Cell differentiation. Evaluation of the MMSC differentiation capacity was carried out after 8 days in anoxia because of quite rapid decline of MMSC viability with exposure extention, as was described above. Spontaneous and induced osteogenic differentiation was revealed in anoxic MMSCs with alkaline phosphatase staining; the intensity of the process was found significantly less pronounced than in normoxic MMSCs (Fig.3). The mechanism of O_2-mediated suppression of osteogenic capacity of human bone marrow MMSCs and murine osteoblasts in the condition of O_2 deprivation (hypoxia (2% O_2) and anoxia (0.02% O_2)) was demonstrated earlier. The authors made a supposition that anoxia rather than hypoxia provoked inhibition of Runx2 protein expression, the key transcription factor in osteogenesis. Runx2 suppression resulted in inhibition of nodule formation and a significant reduction in mineralization of the extracellular matrix [Salim et al., 2004].

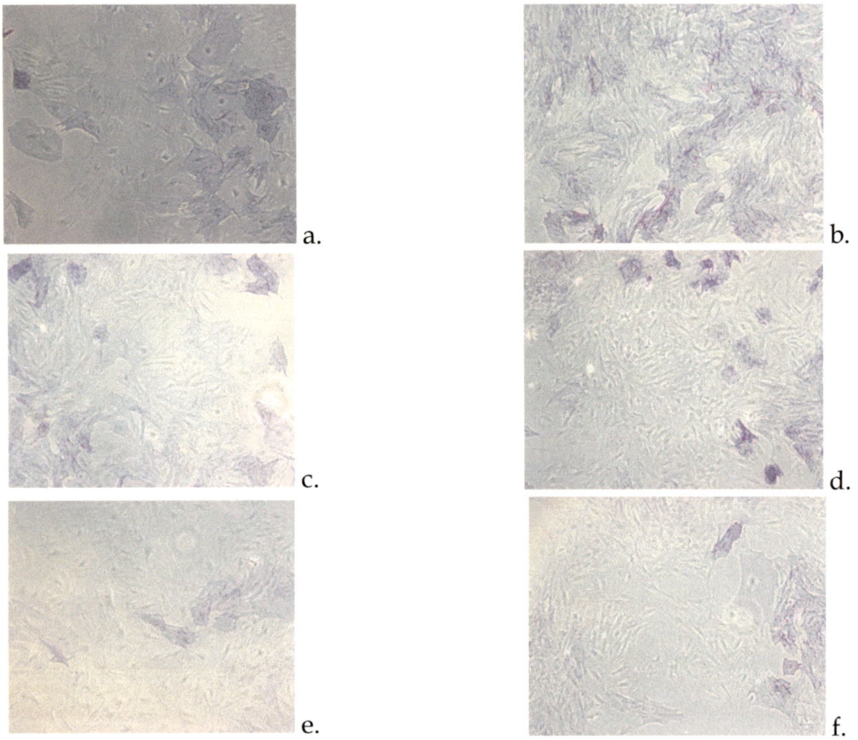

Fig. 3. Spontaneous (a,c,e) and induced (b,d,f) osteodifferentiation in rat bone marrow MMSCs after 8 days exposure in normoxic (20% O_2) (a,b), hypoxic (5% O_2) (c,d) and anoxic (~0% O_2) (e,f) conditions. Alkaline phosphatase, representative images of MMSCs on 3rd passage, 100x.

Accumulation of lipid droplets in anoxic MMSCs indicated of differentiation in the adipogenic direction was both spontaneus and induced (Fig. 4). Further extension of exposure in anoxia caused death of differentiating MMSCs. It appears that alteration of MMSCs viability rather than of differentiating capacity was the cause of differentiation suppression in anoxia.

Thus, 96 h anoxia didn't lead to changes in MMSC morphology, proliferation rate and immunophenotype, which may obviously indicate MMSC functional stability under reduced oxygen tension. Also, anoxia didn't increase significantly the percentage of damaged cells despite some activation of apoptosis. Further MMSC propagation in anoxia led to progressive damage of cells mainly by necrosis in contrast to apoptosis as a main death pathway in short-term anoxia and the antiapoptotic effect of hypoxia. Short-term anoxia did not inhibit the initial stages of stimulated adipo- and osteo-differentiation [Tuncay et al., 1994; Matsuda et al., 1998]. Probably, trace oxygen is enough for some MMSCs to start differentiation and the only limiting factor is viability in anoxia rather than termination of the differentiation signaling pathways.

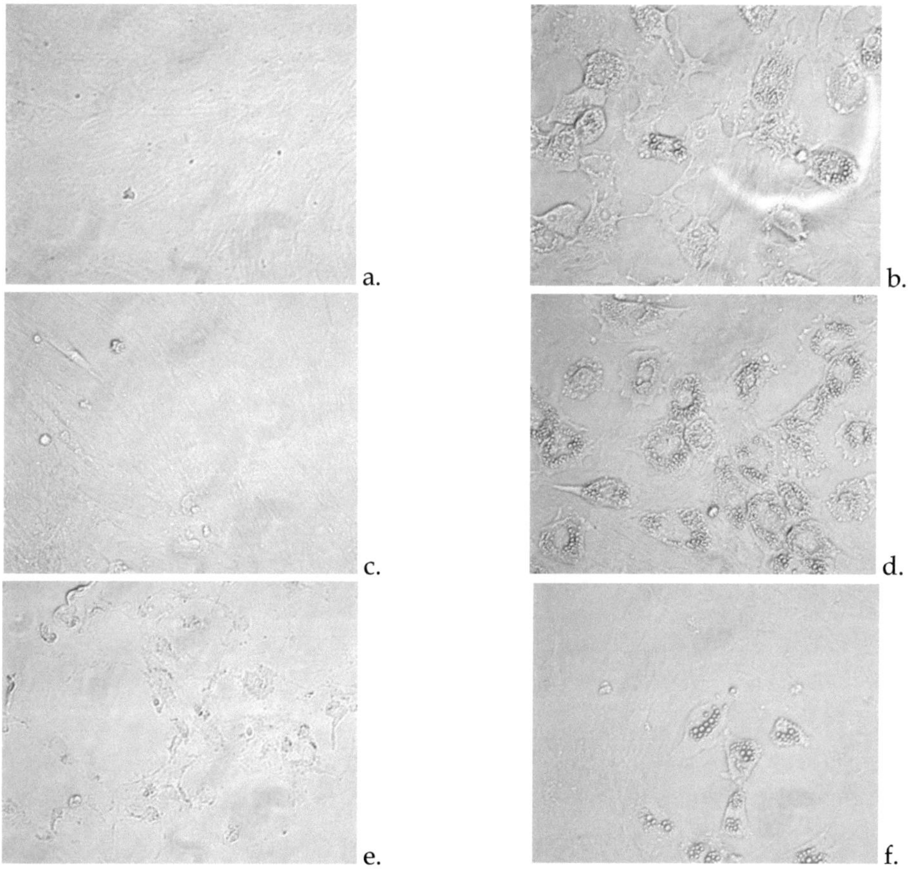

Fig. 4. Spontaneous (a,c,e) and induced (b,d,f) adipodifferentiation in rat bone marrow MMSCs after 4 days exposure in normoxic (20% O_2) (a,b), hypoxic (5% O_2) (c,d) and anoxic (~0% O_2) (e,f) conditions. Lipid droplets were evaluated with Oil Red O staining, representative images of MMSCs on 3rd passage, 400x.

4. Susceptibility of human adipose-tissue derived MMSCs to anoxia

In order to expand our understanding of stromal precursor adaptation to extremely low oxygen we continued studies of anoxia effects on human MMSCs from adipose tissue. MMSCs of 2-4 passsages were subjected to anoxic condition over 240 hours. We did not change culture medium in the course of exposure in anoxia, which induced an additional stress from nutrient "starvation". Cells of the same passage were placed in normoxia (20% O_2) as reference "starvation" cells, and also reference"standard" cells with regular medium replacement every third day were used.

Cell growth. To characterize cell growth, we evaluated increase in MMSC population in each experimental condition. In the "standard" condition at 20% O_2 MMSC population grew 7.1 folds (Fig. 5). After 240 h w/o medium change, i.e. under nutrients deprivation, growth of normoxic MMSCs made up only 4.4 folds. In anoxic cultures cell population increased 3.9 folds (Fig. 5). Against expectations, anoxia did not suppress MMSCs growth. Increase in

MMSC number was slightly less pronounced than in normoxic MMSCs in the condition of starvation. It appears, that starvation decreased oxygen demand but did not stop cell proliferation. The mechanism regulating MMSC proliferation under increasing oxygen limitation is still unclear.

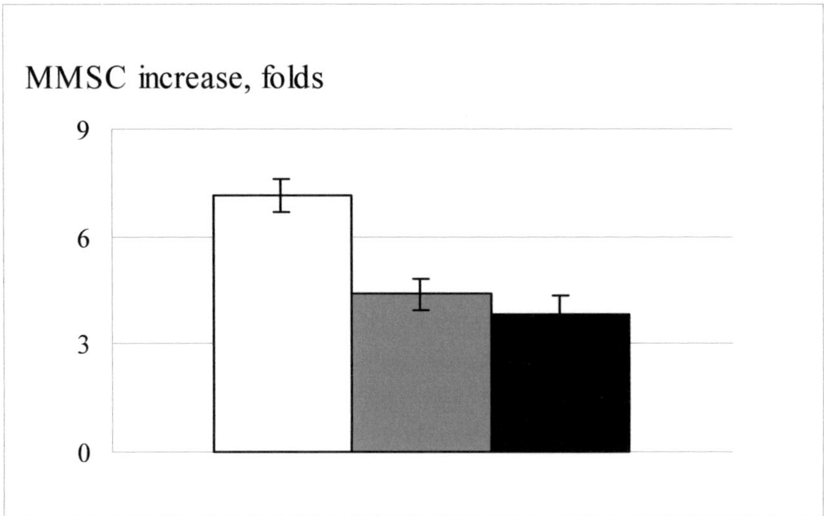

Fig. 5. MMSC increment during expansion in in following conditions: □ - 20% O_2, with regular medium changes (control), -20% O_2, w/o medium replacement and ■ ~0% O_2, w/o medium replacement. Results are representative of three independent experiments. Data are shown as M±m.

Cell viability. We compared MMSC viability in the experimental conditions described in the context of cell death path (Tabl. 4). After 240 h in the standard normoxic condition, the share of necrotic and apoptotic MMSCs was similar and fairly low. Growth of MMSCs in normoxia and anoxia w/o medium replacement shifted the ratio of cell death path toward necrotic. Comparison of these data with MMSC proliferation makes it evident that cell population increase is higher in anoxia and that percent of necrotic cells is same as in normoxia, which means MMSCs growth prevails in anoxia.

	AnnV+ cells (%)	*PI + cells (%)*	*Damaged cells, total (%)*
Standard culture condition			
20% O_2 (control)	2,25 ± 0,12	2,20 ± 0,11	4,45 ± 0,22
Growing cells, 240 hrs without medium replacement			
20% O_2	0,70 ± 0,09	9,17± 0,42	9,87 ± 0,45
20%→0% O_2	2,51 ± 0,03	9,46 ± 0,08	11,97 ± 0,05

The averaged data of 3 independent experiments are presented as M±m.

Table 3. Human adipose-tissue derived MMSC viability after exposure in normoxia and anoxia.

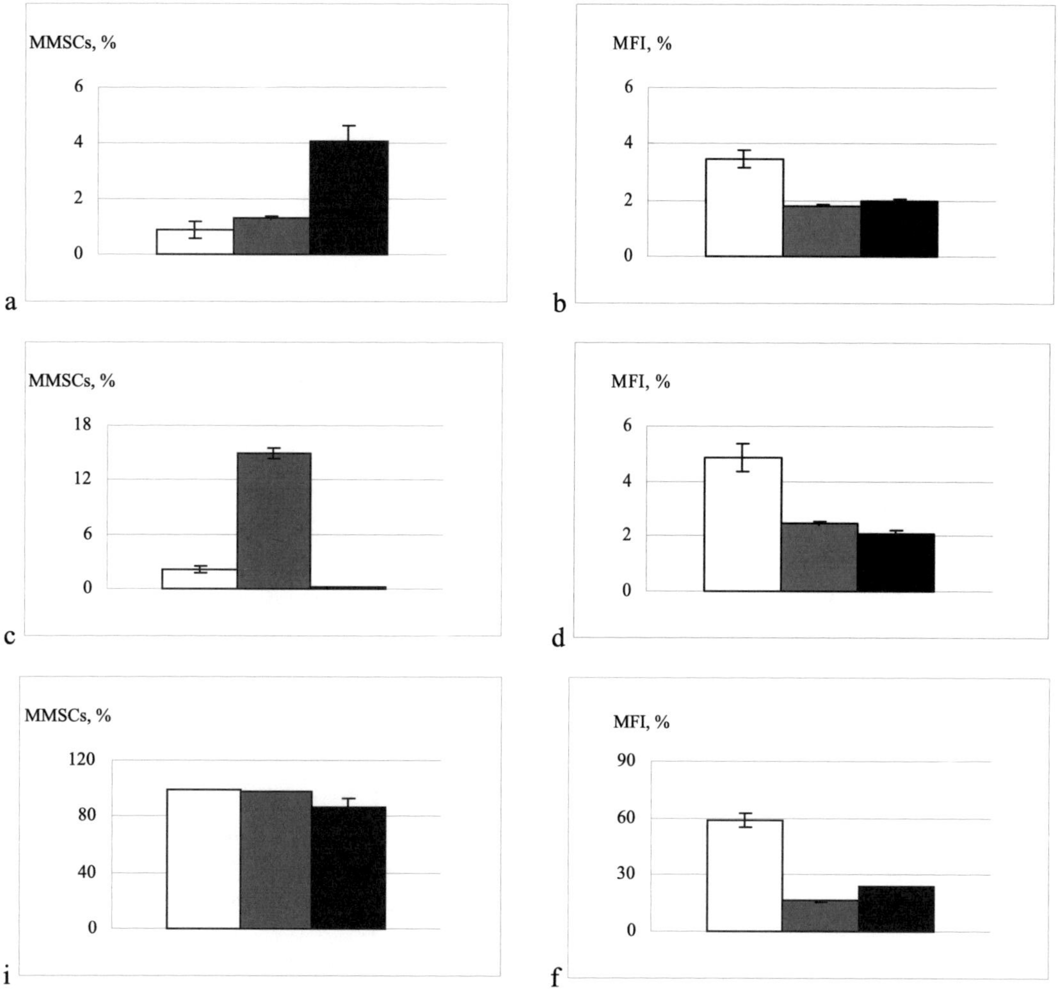

Fig. 6. Cellular metabolism alterations in MMSCs after 240 h exposure in different culture conditions: □ - 20% O_2, with regular medium changes (control), ■ -20% O_2, w/o medium replacement and ■ ~0% O_2, w/o medium replacement. a,b. ROS production was evaluated with H2CFDA; c,d. Lyzosomes were stained with LysoTracker Green, f. Mitochondria were stained with MitoTracker Red.a,c,e - Number of stained MMSCs; b,d,f - Mean fluorescence intensity per cell. Results are representative of three independent experiments. Data are shown as M±m.

Cellular organelle status. To clarify the mechanisms underlining MMSC resistance to extremely low oxygen in microenvironment, we characterized alterations in some vital parameters of cellular homeostasis (Fig. 6). After 240 h of anoxia, MMSCs demonstrated an increase in percent of MMSCs containing intracellular ROS (4% vs 1.2% in normoxia) w/o increase in mean fluorescence intensity (MFI) per a cell (the parameter describes the relative amount of fluorochrome in the cell). The number of MMSCs with active lyzosomes and unchanged MFI decreased drastically in anoxia. It is interesting, that long-term anoxia neither affected the mitochondrial compartment (all cells imposed active mitochondria) nor influenced the mitochondrial transmembrane potential. Thus, in comparison with normoxia, anoxia increased slightly the number of cells with intracellular ROS and decreased significantly the number of MMSCs with active lyzosomes.

It is nessesary to underline that cell growth potential and viability of human adipose-derived MMSCs depend more on nutrient's supply than on the concentration of oxygen. The drastical increase in the number of cells with active lysosomes under combination of nutrients starvation and anoxia should be clarified further.

Taking together, presented data confirm that MMSCs from different source are extraordinary resistant to extermely low oxygen in the microenvironment. In anoxic conditions MMSCs retain their properties to proliferate and differentiate in mesenchyma-specific lineages and also possess fairly high viability. Nevertheless, long-term anoxia provoked cell death in MMSCs mainly through necrotic pathway.

5. Concluding remarks

The data reviewed above and our own results demonstrate clearly that low oxygen tension is undoubtedly an important regulator of MMSCs maintenance and plays a pivotal role in architecturing the MMSCs microenvironment. With oxygen partial pressure sublethal or lethal for other cell types, MMSCs not only survive but enhance proliferative activity and slow down differentive capacity supporting "stemness". Low oxygen conditions accentuate the paracrine role of MMSCs by altering the soluble factor release which is also plays an important role in mobilizing MMSCs and recruiting them to site of injury. MMSC ability to outlive low/extremely low oxygen (anoxia) probably, is based on their capacity to upregulate survival pathways and increase glycolytic metabolism. MMSCs are very tolerant of oxygen starvation keeping their morphology, immunophenotype and proliferation rate, demonstrating slightly affected metabolism and potency to mesenchymal lineage differentiation. Short-term anoxia gives start to apoptosis in contrast to hypoxia which exerts the antiapoptotic effects on MMSCs. Long-term anoxia provokes progressive MMSC damage mainly through the necrotic pathway. The effects of hypoxia and anoxia may be diverse and accounted to the fact, that low oxygen simulates the *in vivo* conditions and can be regarded as an approximation of the physiological MMSC milieu rather than the hypoxic impact. On the contrary, anoxia may represent a real hypoxic microenvironment for these cells.

These results are very encouraging both for understanding particular mechanisms of MMSC existence in different microenvironments and cell therapy as an instrument of MMSC *ex vivo* modification. In view of the outstanding properties, MMSCs are considered as a perspective tool for cell therapy and regenerative medicine. These cells have already shown a great regenerative potential in preclinical studies and clinical trials. The quality of cell product for these purposes is very important. Up to now, in a few studies MMSCs *ex vivo* expanded at low oxygen and anoxia demonstrated regenerative properties superior to MMSCs propagated by the standard cultivation. The possibility to modify MMSC properties *ex vivo* opens great opportunities for implication of 'hypoxic" or even "anoxic" protocol for MMSC expansion to meet the needs of cell therapy. Nevertheless, the question about the true hypoxic environment for mesenchymal stromal progenitor cells still has not got the final answer and invites further investigations.

6. Acknowledgments

The work in the authors' laboratories was supported by grants from the Russian Ministry of High Education and Science #02.522.11.2006, #11.G34.31.0006 and Program of the Department of Biological Science, Russian Academy of Sciences.

7. References

Amellem O, Loffler M, Pettersen EO. Regulation of cell proliferation under extreme and moderate hypoxia: the role of pyrimidine (deoxy) nucleotides. Br J Cancer. 1994;70(5):857-866.

Annabi B, Lee YT, Turcotte S, Naud E, Desrosiers RR, Champagne M, Eliopoulos N, Galipeau J, Béliveau R. Hypoxia promotes murine bone-marrow-derived stromal cell migration and tube formation. Stem Cells. 2003;21(3):337-347.

Anokhina EB, Buravkova LB, Galchuk SV. Resistance of rat bone marrow mesenchymal stromal precursor cells to anoxia in vitro. Bull Exp Biol Med. 2009;148(1):148-151.

Augello A, Kurth TB, De Bari C. Mesenchymal stem cells: a perspective from in vitro cultures to in vivo migration and niches. Eur Cell Mater. 2010;20:121-33.

Betre H, Ong SR, Guilak F, Chilkoti A, Fermor B, Setton LA. Chondrocytic differentiation of human adipose-derived adult stem cells in elastin-like polypeptide. Biomaterials. 2006;27(1):91–99.

Bianco P, Riminucci M, Gronthos S, Robey PG. Bone marrow stromal stem cells: nature, biology, and potential applications. Stem Cells. 2001;19(3):180-192.

Black AC: Human reserve pluripotent mesenchymal stem cells are present in the connective tissues of skeletal muscle and dermis derived from fetal, adult, and geriatric donors. Anat Rec. 2001;264:51-62.

Bosch P, Musgrave DS, Lee JY, Cummins J, Shuler T, Ghivizzani TC, Evans T, Robbins TD, Huard J. Osteoprogenitor cells within skeletal muscle. J Orthop Res. 2000;18:933-944.

Bosch P, Pratt SL, Stice SL. Isolation, characterization, gene modification, and nuclear reprogramming of porcine mesenchymal stem cells.Biol Reprod. 2006;74(1):46-57.

Brahimi-Horn MC, Pouysse'gur J. Oxygen, a source of life and stress. FEBS Lett. 2007;581(19):3582-3591.

Braun RD, Lanzen JL, Snyder SA, Dewhirst MW. Comparison of tumor and normal tissue oxygen tension measurements using OxyLite or microelectrodes in rodents. Am. J. Physiol. Heart Circ. Physiol. 2001;280:H2533–H2544.

Bruder SP, Kraus KH, Goldberg VM, Kadiyala S. The effect of implants loaded with autologous mesenchymal stem cells on the healing of canine segmental bone defects. J Bone Joint Surg. 1998;80(7):985-996.

Buravkova LB, Andreeva ER, Zhambalova AP, Romanov YuA. Hematopoiesis induced microenvironment: the role of oxygen and stromal cells. Aerospace and Ecology Medicine. 2010; 44(5):3-8.

Buravkova LB, Anokhina EB. Effect of hypoxia on stromal precursors from rat bone marrow at the early stage of culturing. Bull Exp Biol Med. 2007;143(4):411-413.

Buravkova LB, Anokhina EB. Mesenchymal stromal progenitor cells: general characteristics and functional state in low oxygen tension. Russian Physiol J. 2008;94(7):737-57.

Buravkova LB, Grinakovskaia OS, Andreeva EP, Zhambalova AP, Kozionova MP. Characteristics of human lipoaspirate-isolated mesenchymal stromal cells cultivated under a lower oxygen tension. Cell Tiss Biol. 2009;3(1):23–28.

Caballero M, Reed CR, Madan G, van Aalst JA. Osteoinduction in umbilical cord- and palate periosteum-derived mesenchymal stem cells. Ann Plast Surg. 2010;64(5):605-609.

Calvi LM, Adams GB, Weibrecht KW, Weber JM, Olson DP, Knight MC, Martin RP, Schipani E, Divieti P, Bringhurst FR, Milner LA, Kronenberg HM, Scadden DT. Osteoblastic cells regulate the haematopoietic stem cell niche. Nature. 2003;425(6960):841-846.

Caplan AI. Adult mesenchymal stem cells for tissue engineering versus regenerative medicine. J Cell Physiol. 2007;213(2):341-347.

Csete M. Oxygen in the cultivation of stem cells. Ann N Y Acad Sci. 2005;1049:1-8.

Cipolleschi MG, Dellosbarba P, Olivotto M. The role of hypoxia in the maintenance of hematopoietic stem-cells. Blood.1993;82(7):2031–2037.

Das R, Jahr H, van Osch GJ, Farrell E. The role of hypoxia in bone marrow-derived mesenchymal stem cells: considerations for regenerative medicine approaches. Tissue Eng Part B Rev. 2010;16(2):159-168.

D'Ippolito G, Diabira S, Howard GA, Menei P, Roos BA, Schiller PC. Marrow-isolated adult multilineage inducible (MIAMI) cells, a unique population of postnatal young and old human cells with extensive expansion and differentiation potential. J Cell Sci. 2004;117(Pt 14):2971–2981.

Eliasson P, Jönsson JI. The hematopoietic stem cell niche: low in oxygen but a nice place to be. J Cell Physiol. 2010;222(1):17-22.

Erecinska, M., and Silver, I.A. (2001). Tissue oxygen tension and brain sensitivity to hypoxia. Respir. Physiol. 128, 263–276.

Fehrer C, Brunauer R, Laschober G, Unterluggauer H, Reitinger S, Kloss F, Gully C, Gassner R, Lepperdinger G. Reduced oxygen tension attenuates differentiation capacity of human mesenchymal stem cells and prolongs their lifespan. Aging Cell. 2007;6(6):745–757.

Friedenstein AJ, Gorskaja JF, Kulagina NN. Fibroblast precursors in normal and irradiated mouse hematopoietic organs. Exp. Hematol. 1976;4(5):267-274.

Gardner LB, Li Q, Park MS, Flanagan WM, Semenza GL, Dang CV. Hypoxia inhibits G1/S transition through regulation of p27 expression. J Biol Chem. 2001;276(11):7919-26.

Gimble JM, Katz AJ, Bunnell BA. Adipose-Derived Stem Cells for Regenerative Medicine. Circ Res. 2007;100(9):1249-1260.

Goda N, Ryan HE, Khadivi B, McNulty W, Rickert RC, Johnson RS. Hypoxia-inducible factor 1alpha is essential for cell cycle arrest during hypoxia. Mol Cell Biol. 2003;23(1):359-69.

Grayson WL, Zhao F, Izadpanah R, Bunnell B, Ma T. Effects of hypoxia on human mesenchymal stem cell expansion and plasticity in 3D constructs. J Cell Physiol. 2006;207(2):331–339.

Greijer AE, van der Wall E. The role of hypoxia inducible factor 1 (HIF-1) in hypoxia induced apoptosis. J Clin Pathol. 2004;57(10):1009-14

Grinakovskaya OS, Andreeva ER, Buravkova LB, Rylova YV, Kosovsky GY. Low level of O2 inhibits commitment of cultured mesenchymal stromal precursor cells from the adipose tissue in response to osteogenic stimuli. Bull Exp Biol Med. 2009;147(6):760-763

He A, Y Jiang, C Gui, Y Sun, J Li, J Wang. The antiapoptotic effect of mesenchymal stem cell transplantation on ischemic myocardium is enhanced by anoxic preconditioning. Can J Cardiol 2009;25(6):353-358.

Hu X, Yu SP, Fraser JL, Lu Z, Ogle ME, Wang JA, Wei L. Transplantation of hypoxia-preconditioned mesenchymal stem cells improves infarcted heart function via enhanced survival of implanted cells and angiogenesis. J Thorac Cardiovasc Surg. 2008;135(4):799-808.

Ivanovic Z. Hypoxia or in situ normoxia: The stem cell paradigm. J Cell Physiol. 2009;219(2):271-275.

Izadpanah R, Trygg C, Patel B, Kriedt C, Dufour J, Gimble JM, Bunnell BA. Biologic properties of mesenchymal stem cells derived from bone marrow and adipose tissue. J. Cell. Biochem. 2006;99(5):1285-1297.

Jiang BH, Semenza GL, Bauer C, Marti HH. Hypoxia-inducible factor 1 levels vary exponentially over a physiologically relevant range of O_2 tension. Am J Physiol. 1996;271(4 Pt 1):C1172-80.

Jones DL, Wagers AJ. No place like home: anatomy and function of the stem cell niche. Nat Rev Mol Cell Biol. 2008;9(1):11-21.

Khan WS, Adesida AB, Hardingham TE. Hypoxic conditions increase hypoxia-inducible transcription factor 2alpha and enhance chondrogenesis in stem cells from the infrapatellar fat pad of osteoarthritis patients. Arthritis Res Ther 2007;9(3):R55.

Kolf CM, Cho E, Tuan RS. Biology of adult mesenchymal stem cells: regulation of niche, self-renewal and differentiation. Arthritis Res. 2007;9(1):204.

Kurose R, Ichinohe S, Tajima G, Horiuchi S, Kurose A, Sawai T, Shimamura T. Characterization of human synovial fluid cells of 26 patients with osteoarthritis knee for cartilage repair therapy. Int J Rheum Dis. 2010;13(1):68-74.

Lee EY, Xia Y, Kim WS, Kim MH, Kim TH, Kim KJ, Park BS, Sung JH. Hypoxia-enhanced wound-healing function of adipose-derived stem cells: increase in stem cell proliferation and up-regulation of VEGF and bFGF. Wound Repair Regen. 2009;17(4):540-7

Lee JH, Kemp DM. Human adipose-derived stem cells display myogenic potential and perturbed function in hypoxic conditions. Biochem Biophys Res Commun. 2006;341(3):882-888.

Lee RH, Kim B, ChoiI, Kim H, Choi HS, Suh K, Bae YC, Jung JS. Characterization and expression analysis of mesenchymal stem cells from human bone marrow and adipose tissue. Cell Physiol Biochem. 2004;14(4-6):311–324.

Leist M, Single B, Castoldi AF, Kuhnle S, Nicotera P. Intracellular adenosine triphosphate (ATP) concentration: a switch in the decision between apoptosis and necrosis. J Exp Med. 1997; 185(8):1481-6.

Lennon DP, Edmison JM, Caplan AI. Cultivation of rat marrow-derived mesenchymal stem cells in reduced oxygen tension: effects on in vitro and in vivo osteochondrogenesis. J Cell Physiol. 2001;187(3):345–355.

Li L, Xie T. Stem cell niche: structure and function. Annu. Rev.Cell Dev. Biol. 2005;21, 605–631.

Lozito TP, Kolf CM,Tuan RM. Microenvironmental regulation of adult mesenchymal stem cells. In: V.K. Rajasekhar, M.C. Vemuri (eds.), Regulatory Networks in Stem Cells, Stem Cell Biology and Regenerative Medicine, Humana Press, 2009:185-210.

Ma T, Grayson WL, Frohlich M, Vunjak-Novakovic G. Hypoxia and stem cell-based engineering of mesenchymal tissues. Biotechnol Prog. 2009;25(1):32–42.

Madonna R, Geng YJ, De Caterina R. Adipose tissue–derived stem cells characterization and potential for cardiovascular repair. Arterioscler Thromb Vasc Biol. 2009;29(11):1723-1729.

Majumdar MK, Banks V, Peluso DP, Morris EA. Isolation, characterization, and chondrogenic potential of human bone marrow-derived multipotential stromal cells. J. Cell Physiol. 2000;185(1):98-106.

Makino S, Fukuda K, Miyoshi S, Konishi F, Kodama H, Pan J, Sano M, Takahashi T, Hori S, Abe H, Hata J, Umezawa A, Ogawa S. Cardiomyocytes can be generated from marrow stromal cells in vitro. J. Clin. Invest. 1999;103(5):697-705.

Malda J, Klein TJ, Upton Z. The roles of hypoxia in the in vitro engineering of tissues. Tissue Eng. 2007;13(9):2153-2156.

Matsuda N, Morita N, Matsuda K, Watanabe M. Proliferation and differentiation of human osteoblastic cells associated with differential activation of MAP kinases in response to epidermal growth factor, hypoxia, and mechanical stress in vitro. Biochem Biophys Res Commun. 1998;249(2):350-4.

Mohyeldin A, Garzo'n-Muvdi T, Quin~ones-Hinojosa A. Oxygen in stem cell biology: a critical component of the stem cell niche. Cell Stem Cell. 2010; 7(2):150-61.

Panchision DM. The role of oxygen in regulating neural stem cells in development and disease. J Cell Physiol. 2009;220:562–568.

Parmar K, Mauch P, Vergilio JA, Sackstein R, Down JD. Distribution of hematopoietic stem cells in the bone marrow according to regional hypoxia. Proc Natl Acad Sci USA. 2007;104(13):5431–5436.

Pittenger MF, Mackay AM, Beck SC, Beck SC, Jaiswal RK, Douglas R, Mosca JD, Moorman MA, Simonetti DW, Craig S, Marshak DR. Multilineage potential of adult human mesenchymal stem cells. Science. 1999;284(5411):143-147.

Rehman J, Traktuev D, Li J, Merfeld-Clauss S, Temm-Grove CJ, Bovenkerk JE, Pell CL, Johnstone BH, Considine RV, March KL. Secretion of angiogenic and antiapoptotic factors by human adipose stromal cells. Circulation. 2004; 109 (10): 1292–1298.

Ren H, Cao Y, Zhao Q, Li J, Zhou C, Liao L, Jia M, Zhao Q, Cai H, Han ZC, Yang R, Chen G, Zhao RC. Proliferation and differentiation of bone marrow stromal cells under hypoxic conditions. Biochem Biophys Res Commun. 2006;347(1):12-21.

Romanov YA, Darevskaya AN, Kabaeva NV, Antonova OA. Optimum conditions for culturing of human bone marrow and adipose tissue mesenchymal precursor cells. Bull Exp Biol Med. 2006;142(4):515-20.

Rosová I, Dao M, Capoccia B, Link D, Nolta JA. Hypoxic preconditioning results in increased motility and improved therapeutic potential of human mesenchymal stem cells. Stem Cells. 2008;26(8):2173-2182.

Salim A, Nacamuli RP, Morgan EF, Giaccia AJ, Longaker MT. Transient changes in oxygen tension inhibit osteogenic differentiation and Runx2 expression in osteoblasts. J Biol Chem. 2004;279(38):40007-40016.

Scadden DT. The stem-cell niche as an entity of action. Nature 2006;441:1075–1079.

Schäffler A, Büchler C. Concise review: adipose tissue-derived stromal cells--basic and clinical implications for novel cell-based therapies. Stem Cells. 2007;25(4):818-827.

Schmaltz C, Hardenbergh PH, Wells A, Fisher DE. Regulation of proliferation-survival decisions during tumor cell hypoxia. Mol Cell Biol. 1998;18(5):2845-54.

Schofield R. The relationship between the spleen colony-forming cell and the haemopoietic stem cell. Blood Cells. 1978; 4:7–25.

Shake JG, Gruber PJ, Baumgartner WA, Senechal G, Meyers J, Redmond JM, Pittenger MF, Martin BJ. Mesenchymal stem cell implantation in a swine myocardial infarct model: Engraftment and functional effects. Ann Thorac Surg. 2002;73(6):1919-1925.

Shi S, Grontos S. Perivascular niche of postnatal mesenchymal stem cells in human bone marrow and dental pulp. J Bone Miner Res. 2003;18(4):696–704.

Silvan U, Diez-Torre A, Arluzea J, Andrade R, Silio M, Arechaga J. Hypoxia and pluripotency in embryonic and embryonal carcinoma stem cell biology. Differentiation 2009;78:159–168.

Suzuki H, Tomida A, Tsuruo T. Dephosphorylated hypoxia-inducible factor 1alpha as a mediator of p53-dependent apoptosis during hypoxia. Oncogene. 2001;20(41):5779-88.

Tuncay OC, Barker MK Oxygen tension regulates osteoblast function. Am J Orthod Dentofacial Orthop. 1994; 105(5): 457 - 463.

Valtieri M, Sorrentino A. The mesenchymal stromal cell contribution to homeostasis. J. Cell Physiol. 2008;217(2):296-300.

Villarruel SM, Boehm CA, Pennington M, Bryan JA, Powell KA, Muschler GF. The effect of oxygen tension on the in vitro assay of human osteoblastic connective tissue progenitor cells. J Orthop Res. 2008;26(10):1390-7.

Vlaski M, Lafarge X, Chevaleyre J, Duchez P, Boiron JM, Ivanovic Z. Low oxygen concentration as a general physiologic regulator of erythropoiesis beyond the EPO-related downstream tuning and a tool for the optimization of red blood cell production ex vivo. Exp Hematol. 2009;37(5):573-584.

Wagner W, Wein F, Seckinger A, Frankhauser M, Wirkner U, Krause U, Blake J, Schwager C, Eckstein V, Ansorge W, Ho AD. Comparative characteristics of mesenchymal stem cells from human bone marrow, adipose tissue, and umbilical cord blood. Exp Hematol. 2005;33(11):1402-1416.

Wang DW, Fermor B, Gimble JM, Awad HA, Guilak F. Influence of oxygen on the proliferation and metabolism of adipose derived adult stem cells. J Cell Physiol. 2005;204(1):184–191.

Yin, T., and Li, L. (2006). The stem cell niches in bone. J. Clin. Invest. 116, 1195–1201.

Zhambalova AP, Darevskaya AN, Kabaeva NV, Romanov YA, Buravkova LB. Specific interaction of cultured human mesenchymal and hematopoietic stem cells under conditions of reduced oxygen content. Bull Exp Biol Med. 2009;147(4):525-30.

Zuk PA, Zhu M, Ashjian P, De Ugarte DA, Huang JI, Mizuno H, Alfonso ZC, Fraser JK, Benhaim P, Hedrick MH. Human adipose tissue is a source of multipotent stem cell. Mol Biol Cell. 2002;13(12):4279-4295.

Zuk PA, Zhu M, Mizuno H, Huang J, Futrell JW, Katz AJ, Benhaim P, Lorenz HP, Hedrick MH. Multilineage cells from human adipose tissue: implications for cell-based therapies. Tissue Eng. 2001;7(2):211-228.

Zvaifler NJ, Marinova-Mutafchieva L, Adams G, Edwards CJ, Moss J, Burger JA, Maini RN: Mesenchymal precursor cells in the blood of normal individuals. Arthritis Res. 2000; 2:477-488.

4

The Protective Role of Hypoxic Preconditioning in CNS

Li-Ying Wu, Ling-Ling Zhu and Ming Fan
Institute of Basic Medical Sciences, Beijing, China

1. Introduction

The phenomenon that hypoxic preconditioning (HP) protects against subsequent severer anoxia was discovered approximately two decades ago. Subsequently, the effects of HP have been studied intensively in the whole brain, as well as in living hippocampal or cortical slices, and in *in vitro* cell cultures using various hypoxic model systems [1,8,10,12,14,18,28,36,53,56,67,93,94]. Although the exact mechanisms are not completely disclosed, the underlying molecular mechanisms have been postulated. For example, HP activates a great variety of endogenous protective mediators and/or inhibits amounts of harmful mediators, which combined attenuates a burst of free radicals and ultimately increases the capability of cell survival under severe oxygen deprivation [65,74,93,94].

The central nervous system (CNS) is extremely sensitive to a decrease of oxygen content, due to its high intrinsic rate of oxygen consumption [55]. O_2 consumption in brain is maximum in all organs but with a lesser weight ratio. Moreover, the repair ability of CNS is weak to multiple injuries. So, prevention is more important than treatment in CNS. The hypoxic diseases in brain mainly includes stoke, cerebral palsy, etc. Until now, there are no any effective drugs to protect brain from these diseases. Disclosure of the mechanism of HP will contribute to drug discovery for prevention against hypoxic diseases.

A number of cellular adaptive responses to hypoxia are mediated by a key transcription factor termed hypoxia-inducible factor 1 (HIF-1). Activation of HIF-1 by HP enhances the capability to tolerate severe anoxia or ischemia. The target genes of HIF-1, on the one hand, are involved in energy homeostasis, such as erythropoietin (EPO) in the regulation of erythropoiesis [7,29,30,54,71,89], vascular endothelial growth factor (VEGF) in angiogenesis [6,7,88], glucose transmitters (GLUTs) in glucose uptake [62,96] and glycolytic enzymes of anaerobic glycolysis [5,40,80,82], and on the other hand, in redox homeostasis, such as Bcl-2 and adenovirus E1B 19 kDa-interacting protein 3/BNIP3-like (BNIP3/BNIP3L) and microRNA miR-210 in reduction of reactive oxygen species (ROS) [24,43,60,79,97].

ROS are burst from mitochondria during anoxia owing to the lack of O_2 as a final electron acceptor. A mass of ROS are normally regarded as toxic substances due to their strong aggressivity to biological macromolecule, such as proteins, lipids, DNA, and even organelles [52,57]. The macromolecule or organelles damaged by ROS confer oxidative injury, and ultimately result in cell death. In this regard, reduction of ROS during anoxia/ischemia should lead to the protection of cells. However, a moderate amount of ROS is a stimulus to stabilize HIF-1α [16,69,72,84], a functional subunit of HIF-1, which degrades in normoxia. The duplex roles of ROS will be discussed in this chapter.

During anoxia, a change of cell volume was seldom observed by researchers, however, we noticed the phenomenon and found for the first time that cell volume could be regulated by HP, which caused the protection against severe anoxia. Further analysis showed that sorbitol might mediate the protection [95].

This chapter will review the neuroprotective effects of HP and elaborate the above mentioned mechanisms of HP against anoxia.

2. Activation of HIF-1 by HP and its neuroprotective role in ischemic or anoxic damage

2.1 Characteristic of HIF-1

HIF-1 acts as a pivotal mediator in adaptive responses to hypoxia. It is a heterodimeric transcription factor consisting of an oxygen-regulated HIF-1α subunit and a constitutively expressed HIF-1β subunit. Under normoxia, prolyl sites at 402 and 564 residues of HIF-1α are hydroxylated by proline hydroxylases (PHDs, also known as HIF prolyl hydroxylases, HPH and Egg-laying deficient nine, EGLN) utilizing O_2 and α-ketoglutarate as substrates, Fe^{2+} and Vc as cofactors. Hydroxylated HIF-1α is recognized by a tumor suppressor protein Von Hippel Lindau (VHL), an E3 ubiquitin ligase complex, and then HIF-1α is polyubiquitinated by ubiquitins and hence marked for destruction. Ultimately, HIF-1α is degraded in proteasome by hydrolases [17,21,41,42,72,73,91]. In contrast, 803 site of a conserved asparagine residue in HIF-1α is hydroxylated by factor inhibiting HIF-1 (FIH-1), which blocks the binding of the transcriptional co-activator p300/CBP to the site in HIF-1α and leads to the obstruction of transcriptional activation of HIF-1α. Under hypoxia, hydroxylase activity is inhibited due to lack of O_2 as substrate. On the one hand, 402 and 564 sites of prolyl residues cannot be hydroxylated by PHDs, and therefore HIF-1α cannot be degraded through ubiquitin-proteasome pathway. On the other hand, 803 site of asparagine residue cannot be hydroxylated by FIH-1, and hence HIF-1α has the function of transcriptional activation [35,49,59,70,86].

When the stabilized HIF-1α translocates to nucleus, it heterodimerizes with HIF-1β to form a heterodimeric transcription factor complex HIF-1. Then HIF-1 binds to a conserved cis-regulatory motif called the hypoxia-response element (HRE) that contains the core site of 5'-(A/G)CGTG-3' on its target genes. By interaction with coactivators that are required for transcriptional responses to hypoxia, more than two hundred genes are transactivated by HIF-1. HIF-1 target genes include growth and survival factors, such as VEGF, EPO, heme oxygenase 1 (HO-1) [4,63,99], adrenomedullin (AM) [58], inducible nitric oxide synthase (iNOS) [51], etc.; glucose metabolism, such as glucose transporters GLUT1, GLUT3, phosphoglycerate kinase 1, pyruvate kinase M, lactate dehydrogenase A [78,96], etc.; molecules stabilizing homeostasis of redox, such as BNIP3/BNIP3L [79,97], miRNAs [24], etc., which allow cells to adapt to the hypoxic environment.

In addition to stabilizing HIF-1α protein, hypoxia also leads to increased transcription of HIF-1α mRNA and increased expression of HIF-1α protein in brain [82]. Briefly, hypoxic stimuli play important roles in the accumulation of HIF-1α protein and in transcriptional activation of HIF-1. It has been found that the accumulation of HIF-1α in tissues or cells promoted adaptive mechanisms for cell survival and mediated the tolerance induced by HP [1,87].

2.2 The vital role of HIF-1 in HP

Hypoxic preconditioning (HP) is an exogenous phenomenon in which brief episodes of a hypoxic sublethal insult induces protection against the deleterious effects of subsequent lethal anoxia or ischemia [39,87]. The brain is one of the first organs to fail in hypoxia due to its high intrinsic rate of oxygen consumption. Kitagawa et al. first reported the cerebral HP in 1990 [45]. HP (3 h at 8%O_2) for 24 h before a unilateral occlusion of the common carotid artery in the neonatal rat brain promoted ischemic tolerance [27]. Miller and Bernaudin separately used two different models of hypoxia-induced ischemic tolerance in the adult mouse brain [6,61]. And Bernaudin demonstrated that the tolerance was in association with an increased expression of HIF-1. However, there is no study shown that the effects of hypoxic preconditioning are interrupted by deficiency or blockage of HIF-1 because systemic disruption of the HIF-1 gene leads to embryonic lethality until conditional knockout HIF-1 mice are created. Taie et al. used neural cell-specific HIF-1α-deficient mice to elucidate the role of HIF-1α in hypoxic preconditioning in the brain [87]. . Their results indicated that the protective effects of HP were partially mediated by improving tissue oxygenation via HIF-1α. Similarly, taking advantage of the Cre/Lox technology to generate conditional mutant mice with deletion of HIF-1α predominantly in neurons of the forebrain, Baranova provided evidence that these mutant mice subjected to transient focal cerebral ischemia induced by middle cerebral artery occlusion (MCAO) deteriorated ischemic brain damage. To verify the beneficial role of HIF-1 in the ischemic brain, mice were treated with pharmacologic HIF-1 activators and found that severity of brain ischemic damage was lessened [2]. These researches indicates that activation of HIF-1 by HP is essential to ischemic or anoxic tolerance. However, in contrast to the above results, late-stage brain-specific knock-out of HIF-1α in adult mice reduces rather than increases hypoxic-ischemic damage, and the data suggest that in acute hypoxia, the neuroprotection found in the HIF-1α-deficient mice is mechanistically consistent with a predominant role of HIF-1α as proapoptotic and loss of function leads to neuroprotection [34].

In addition, the stabilization of HIF-1α by HP has been shown to be dose-dependent. Preconditioning of mice dose-dependently stabilized HIF-1α in the retina: exposure to 6 and 10% oxygen for 6 h had the most profound effect on HIF-1α stabilization, while exposure to 14% oxygen stabilized HIF-1α to intermediate levels, and 18 and 21% (normoxia) conditions did not cause a notable stabilization of this transcription factor. Hypoxic HIF-1α stabilization was transient as reoxygenation for 1 h was sufficient to result in the complete degradation of the protein [29].

2.3 The important target genes of HIF-1

Being a transcription factor, HIF-1 plays its important roles via activation of its target genes rather than by itself. The increased expression of HIF-1 by HP results in transcriptional activation of a number of target genes involved in erythropoiesis, angiogenesis, vasodilation, glucose transport, anaerobic glycolysis and autophagy [40,77,79,81,90]. These target genes are the effectors carrying out the protection from severe anoxia or ischemia, whereas HIF-1 is the sensor or executor of the response to hypoxia. Therefore, activation of HIF-1 and its target genes by HP are equal importance to adaptive mechanisms underlying the prevention against ischemic or anoxic damage. By oligonucleotide microarrays to examine genomic responses in neonatal rat brain following 3 h of hypoxia (8% O_2) and either 0, 6, 18, or 24 h of re-oxygenation, Bernaudin showed that 12 HIF-1 target genes including EPO, VEGF , GLUT-1, AM, etc. were involved in brain hypoxia-induced tolerance. The results suggested that HP-induced HIF-1 target genes might mediate neuroprotection against subsequent ischemia, and might provide novel therapeutic targets for treatment of cerebral ischemia [7].

2.4 EPO

The main role of EPO is the stimulation of erythrocyte production. Studies of human and rat brain show that cerebral EPO is produced by both neuronal and glial cells and that neurons, glia, and cerebral endothelial cells all express the EPO receptor [33]. In mice focal cerebral ischemia model, Leconte et al. demonstrated that a late application of hypoxia 5 days after ischemia reduced delayed thalamic atrophy. Further, with an in vitro oxygen glucose deprivation (OGD) model, they found that HIF-1α and the target gene EPO were both increased by hypoxic postconditioning and revealed that EPO was involved in hypoxia postconditioning-induced neuroprotection [50]. In another report, Bernaudin showed that HP with 8% O_2 of 1-hour, 3-hour, or 6-hour duration for 24 hours before ischemia reduced infarct volume caused by focal permanent ischemia in adult mouse brain by approximately 30% when compared with controls, and they also demonstrated that HP rapidly increased the nuclear content of HIF-1α as well as the mRNA level of EPO and its protein level was up-regulated 24 hours after 6 hours of HP. Therefore, they proposed that HIF-1 target genes contributed to the establishment of ischemic tolerance [6]. A report from Prass et al. demonstrated that HP for 180 or 300 minutes induced relative tolerance to transient focal cerebral ischemia, as evidenced by a reduction of infarct volumes to 75% or 54% of the control, respectively, and they found a marked activation of HIF-1 DNA-binding activity and a 7-fold induction of EPO transcription. Infusion of soluble EPO receptor that neutralizes EPO significantly reduced the protective effect of hypoxic pretreatment by 40%. This study was the first to present functional evidence that EPO is an essential mediator of protection in HP. Accordingly, the authors concluded that endogenously produced EPO is an essential mediator of ischemic preconditioning [71]. Recently it has been reported that HP increased secretion of EPO and up-regulated expression of HIF-1α, B cell lymphoma/lewkmia-2 (Bcl-2), erythropoietin receptor (EPOR), neurofilament (NF), and synaptophysin in ES cell-derived neural progenitor cells (ES-NPCs). Interestingly, HP-primed ES-NPCs survived better 3 days after transplantation into the ischemic brain (30-40% reduction in cell death and caspase-3 activation), and transplanted HP-primed ES-NPCs exhibited extensive neuronal differentiation in the ischemic brain, accelerated and enhanced recovery of sensorimotor function when compared to transplantation of non-HP-treated ES-NPCs [89]. The increased secretion of EPO by HP may play the crucial role in the ischemic brain. It has also been reported that EPO acts at EPO receptors to activate Janus kinase-2 (Jak2), which initiates phosphorylation of inhibitor of NF-κB (IκB) to activate nuclear factor-κB

(NF-κB) and induce NF-κB neuroprotective genes [20]. It is another pathway of EPO in neuroprotection.

2.5 VEGF

VEGF is expressed both in endothelial cells and in neural cells (neurons, astrocytes and microglia). It has the ability to promote cerebral angiogenesis and vasodilation. Induction of VEGF by HP is an attempt to increase tissue oxygen levels by improving blood circulation through the formation of new vessels. Therefore, endogenous VEGF over-expression stimulated by HP should be beneficial following stroke by improving oxygen and nutrient delivery to the ischemic area. In the study of ischemic tolerance, it has been reported that HP with normobaric hypoxia induced ischemic tolerance in adult mice, and increased the nuclear content of HIF-1α as well as the mRNA and protein levels of VEGF [6]. Although the authors did not provide a direct cause-relation effect of VEGF on tolerance induction, the increased expression of VEGF at the time of tolerance appearance provided an indirect argument speaking for the possible implication of HIF-1α and its target genes in this phenomenon. In contrast, the $VEGF^{\partial/\partial}$ knock-in mice, which lack the HRE in the VEGF promoter, reduced hypoxic VEGF expression and caused motoneuron degeneration. This finding suggests an important role for VEGF in neuronal development and maintenance within the central nervous system [48,68].

In the retina, VEGF is also recognized as a pro-survival factor protecting retinal neurons against ischemic injury. Nishijima et al. demonstrated that ischemic preconditioning 24 hours before ischemia-reperfusion injury increased VEGF-A (also called VEGF) levels and substantially decreased the number of apoptotic retinal cells. The protective effect of ischemic preconditioning was reversed after VEGF-A inhibition. Thus, they hypothesized that treatment with VEGF-A might provide neuroprotection in the retina, particularly during ischemic eye disease [64]. Cerebellar granule neurons exposed to 5% O_2 for 9 h showed increased levels of VEGF, VEGF receptor-2 (VEGFR-2), phosphorylated Akt/protein kinase B (PKB), and extracellular signal-regulated kinase 1 (ERK1). Incubation with a neutralizing anti-VEGF antibody, a monoclonal antibody to VEGFR-2, wortmannin, or antisense-Akt/PKB, an ERK-inhibitor, reversed the resistance acquired by HP. Inhibition of VEGFR-2 blocked the activation of Akt/PKB. Pretreatment with recombinant VEGF resulted in a hypoxia-resistant phenotype in the absence of HP. These data indicate a requirement for VEGF/VEGFR-2 activation for neuronal survival mediated by HP and suggest VEGF as a hypoxia-induced neurotrophic factor [92]. The above findings implicate the potential use of VEGF as a therapeutic in neural ischemic or anoxic diseases.

2.6 HO-1

Heme oxygenase (HO) belongs to the heat-shock protein families. It is the rate-limiting enzyme for oxidizing heme to biliverdin and carbon monoxide. Biliverdin is further metabolized to bilirubin, which is a strong antioxidant. By means of guanylyl cyclase, carbon monoxide works as an intracellular messenger, similar to nitric oxide [4]. The isoform heme oxygenase 1 (HO-1; also called HSP32) was found to be inducible in a variety of stress conditions, such as hypoxia [25,47], heat shock [66], hydrogen peroxide [38,55] and so on. The induction is considered to be a cellular adaptive protection due to the antioxidation of HO-1[4,38,55,63,98].

HO-1 is a HIF-1 target gene that has been shown to be expressed in the cerebellum following focal ischemia and in the retina following repetitive HP [33]. Because ischemia or anoxia have been associated with increased ROS, clearance of ROS by production of antioxidants by HO-l could help protect the brain from oxidative injury. As a downstream gene of HIF-1, HO-1 can be induced by HP. Garnier et al. demonstrated that HP induced a progressive and sustained expression of HO-1, and they got a conclusion that antioxidant enzymatic defenses in response to hypoxia might be involved in the protective effect of HP against hypoxia–ischemia [25]. Similarly, in rat liver HO-1 is also involved in the protection exerted by HP against hepatic ischemia-reperfusion (I/R) injury. Lai et al. showed that the levels of HO-1 mRNA and protein were obviously over-expressed after 2 weeks of HP. HP diminished the injury after I/R, while after inhibition of HO-1 activity by zinc protoporphyrin (ZnPP), the protective effect of HP was lessened [47]. Furthermore, they demonstrated that pharmacological preconditioning with simvastatin protected liver from I/R injury by HO-1 induction [46]. In contrast, suppressing HO-1 expression in the presence of HO-1 siRNA during I/R injury, apoptosis was enhanced, whereas HO-1 over-expression attenuated apoptosis [55]. It suggests the anti-apoptosis activity of HO-1.

2.7 BNIP3

BNIP3 (Bcl-2/adenovirus E1B 19-kDa interacting protein 3), BH3-only Bcl-2 family member, localizes to mitochondria when overexpressed [9]. It positively regulates autophagy by competing with beclin-1, a highly conserved protein that is required for the initiation of autophagy, for binding to Bcl-2 or Bcl-X_L, thereby releasing beclin-1 to induce autophagy. Activation of BNIP3-dependent autophagy decreases mitochondrial mass and ROS formation [3,97].

BNIP3 is a known HIF-1 target gene. The promoter has two HIF-1 binding sites. BNIP3 is suppressed by Von Hippel-Lindau (VHL) protein in a renal cell carcinoma cell line, consistent with its regulation through the HIF-1α pathway [85]. In an ischemia-reperfusion model, BNIP3 induced autophagy, which protected myocytes from cell death [31]. Zhang et al. showed that hypoxia induced mitochondrial autophagy and this process required the HIF-1-dependent expression of BNIP3 and the constitutive expression of Beclin-1. The authors proposed a molecular pathway: prolonged hypoxia stimulates HIF-1α activity that up-regulates BNIP3 expression, then over-expression of BNIP3 disrupts the interaction of Beclin-1 with Bcl-2, and as a result, BNIP3 competitively binds to Bcl-2, which leads to Beclin-1 release. Subsequently, Beclin-1 recruits autophagy related proteins that induce the occurrence of mitochondrial autophagy (also called mitophagy). After the defective mitochondria are cleared by mitophagy, cell survival is enhanced under hypoxia due to the reduction of ROS by mitophagy.

In addition, on the basis of the characteristics of HIF-1, pharmacological stabilization or activation of HIF-1α will contribute to the protection against ischemia or severe anoxia, imitating the effect of HP. Desferrioxamine (DFO) and cobalt chloride ($CoCl_2$), an iron chelator and competitive inhibitor of iron, respectively, usually used as a positive control are extensively used agents to mimic the effect of HP by inhibiting PHD enzyme activity and thus stabilize HIF-1α [33,55,78,91]. Additionally, the 2-OG analogues L-mimosine (L-mim), dimethyloxalylglycine (DMOG), and 3,4-dihydroxybenzoate (3,4-DHB) can also be used to inhibit PHD enzyme activity and stabilize HIF-1α [33,71]. Recently it has been shown that a PHD inhibitor, 2,4-pyridinedicarboxylic acid diethyl ester, pretreatment followed by a 30-min oxygen-glucose deprivation enhanced neuronal resistance in organotypic hippocampal slices on a model of ischemic damage, which was similar to the effect of anoxia preconditioning in the same model system [56]. With the development of novel PHD inhibitors for the treatment of ischemic diseases, a clinical treatment will be promising.

Taken together, these observations suggest that HIF-1 is involved in mediating the beneficial effects of preconditioning, and that pharmacological activation of HIF-1 may be beneficial in stroke and other hypoxic diseases.

3. The duplex roles of ROS

3.1 Formation of ROS

ROS are small and highly reactive molecules that can damage proteins, lipids, DNA and even mitochondria. In mitochondria, respiration generates ROS and ROS are constantly produced during normal metabolism. The balance of ROS formation and removal of ROS via enzymatic activity or antioxidants establishes constitutive ROS levels. ROS include superoxide anions ($O^{2-\bullet}$), hydrogen peroxide (H_2O_2), hydroxyl radical ($^\bullet OH$). $O^{2-\bullet}$ are the most common ROS, but can be readily converted to other types by enzymatic (e.g. superoxide dismutase) and non-enzymatic reactions. $O^{2-\bullet}$ are formed by metabolic processes (especially in mitochondria) or by enzymatic reactions (e.g. NADPH oxidase). Multiple isoforms of superoxide dismutase (SOD) can catalyze $O^{2-\bullet}$ to form H_2O_2. $^\bullet OH$ is formed from the breakdown of H_2O_2 by transition metals such as iron (Fe^{2+}) or copper (Cu^+) through *Fenton reactions*; this is a particularly reactive form of ROS with a short half life. $^\bullet OH$, in turn, can form from $O^{2-\bullet}$ and H_2O_2 via the *Haber-Weiss reaction* [100].

Hypoxia and re-oxygenation are capable of stimulating ROS formation in brain tissue. Hypoxia stimulates ROS formation from mitochondria and xanthine oxidase in the cortex, whereas reoxygenation induces NADPH oxidase-derived ROS formation. Although single bouts of hypoxia or re-oxygenation increase ROS formation, repetitive hypoxia/re-oxygenation events amplify this effect. Greater ROS formation during and/or following repeated hypoxia and re-oxygenation may contribute to the pattern-sensitivity of hypoxia-induced respiratory plasticity [57].

3.2 Damage roles of ROS and rescue by autophagy

Excess ROS can damage DNA and proteins, contributing to aging, cardiac disease, cancer and other pathologies [23]. The damage can induce the mitochondrial permeability transition (MPT) caused by opening of non-specific high conductance permeability transition (PT) pores in the mitochondrial inner membrane. ATP depletion from uncoupling of oxidative phosphorylation then promotes necrotic cell death, whereas release of cytochrome c after mitochondrial swelling activates caspases and onset of apoptotic cell death. The defective mitochondria have the potential for futile ATP hydrolysis, which accelerates production of ROS and release of proapoptotic proteins. ROS also attack nucleic acids and are thus genotoxic. A lack of histones in mitochondrial DNA (mtDNA) accounts, at least in part, for a 10- to 20-fold higher mutation rate of mtDNA compared to nuclear

DNA [44]. It is thus clear that elimination of dysfunctional mitochondria is essential to protect cells from the injury of disordered mitochondrial metabolism.

Cells can activate a range of pathways to eliminate ROS, or modulate ROS levels to facilitate essential cellular repair [101]. Mitochondrial autophagy has recently been described as an adaptive metabolic response to prevent increased levels of ROS and cell death [97]. It is now widely accepted that autophagy is crucial for removal of damaged mitochondria by mitophagy. Activation of BNIP3-dependent mitophagy decreases mitochondrial mass and ROS formation, thus supporting cell survival under hypoxia [102]. Removal of dysfunctional mitochondria or oxidized proteins by autophagy is suggested to take place via the chaperone-mediated autophagy (CMA) pathway [103]. By interacting with LC3, p62 mediates the targeting of damaged mitochondria into autophagosomes. More recently, p62 has been implicated in the delivery of oxidized proteins to autophagosomes for degradation. Initially, p62 was thought to act solely through the ubiquitin-proteasome system. However, growing data demonstrates the crucial role of p62 in delivering oxidized protein aggregates to autophagosomes [75]. Removal of damaged mitochondria and oxidized proteins, in most cases, supports survival. Therefore, autophagy is primarily a survival mechanism in response to ROS.

In addition to autophagy, HIF-1 pathway is also involved in decreasing ROS levels. HIF-1 reduces ROS production under hypoxic conditions by multiple mechanisms including: a subunit switch in cytochrome c oxidase from the COX4-1 to COX4-2 regulatory subunit that increases the efficiency of complex IV [104]; induction of pyruvate dehydrogenase kinase 1, which shunts pyruvate away from the mitochondria [105]; and induction of microRNA-210, which blocks expression of the iron-sulfur cluster assembly proteins ISCU1/2 that are required for the function of the tricarboxylic acid (TCA) cycle enzyme aconitase and electron transport chain (ETC) complex I [79]. These HIF-1-mediated mechanisms in maintaining redox homeostasis suggest an adaptive response to hypoxia.

3.3 Beneficial roles of ROS through stabilizing HIF-1

In contrast to the potential damaging effects of ROS, recent studies suggest that generation of mitochondrial ROS, especially H_2O_2, is required for hypoxic HIF-1 activation and stabilization [22,32,54]. In human lung epithelial A549 cells, over-expression of antioxidant enzymes that scavenge H_2O_2, such as catalase or glutathione peroxidase 1 (GPx 1) prevents hypoxic stabilization of the HIF-1α protein. However, over-expression of superoxide dismutase 1 or 2, which detoxifies superoxide to H_2O_2, does not alter hypoxic HIF-1 stabililization [11]. Furthermore, low dose exogenous H_2O_2 could trigger HIF-1α expression and thereby, contribute to hypoxic/ischemic (H/I) preconditioning protection in the immature brain. Chang et al. reported that H_2O_2 induced HIF-1 α protein expression in a dose-dependent manner and provided neuroprotection against severe oxygen-glucose deprivation (OGD) 24 h later. They observed that low dose of exogenous H_2O_2 not only conferred cells a tolerance to subsequent lethal insult, but also alone significantly up-regulated HIF-1α protein expression, suggesting that H_2O_2 produced during OGD preconditioning may stabilize and upregulate HIF-1α. HIF-1α stabilized by endogenous H_2O_2 induced by H/I preconditioning might mediate H/I preconditioning protection [15]. Using superoxide dismutase (SOD1) transgenic (Tg) mice, Liu et al. found that hypoxic preconditioning (HP) was protective in wild-type (Wt) neurons but not in neurons obtained from SOD1 Tg mice [54]. In Wt neurons, HIF-1α and EPO expression showed a greater increase after hypoxia compared with Tg neurons. Therefore, the authors concluded that HP induced ROS, which might downregulate the threshold for production of HIF-1α and EPO expression during subsequent lethal hypoxia, thus exerted neuroprotection [54].

In conclusion, ROS plays completely opposite roles depending on the amount and the time of ROS production: less ROS and earlier generation has beneficial roles; conversely, excess ROS and later generation has damaging effects.

4. Regulation of cell volume

Cell volume can be regulated by HP to prevent severe anoxic injury. We demonstrated for the first time that HP protected PC12 cells against necrosis after exposure to acute anoxia (AA), and this protective role of HP was probably related to cell volume regulation by increasing aldose reductase (AR) and sorbitol levels [95].

AR is the first enzyme of the polyol pathway. In this pathway, AR catalyzes conversion of glucose to sorbitol in the presence of nicotinamide adenine dinucleotide phosphate, while sorbitol dehydrogenase converts sorbitol to fructose in the presence of NAD^+ [37]. Under osmotic stress, the abundance of AR is elevated by increasing the transcription of its gene. In turn, sorbitol synthesis is raised by increasing the amount and activity of AR [13]. A change of osmolarity causes sorbitol to leak rapidly to the external medium through a sorbitol permease transport pathway, which prevents excessive cell swelling. Efflux of sorbitol was the primary mechanism for regulatory volume decrease (RVD). RVD protects the cells by minimizing swelling [26].

Sorbitol synthesized from glucose catalyzed by AR is directly related to cell volume regulation [19,76,83]. After synthesis, sorbitol is packed into the secretory vesicles to carry and fuse to cytoplasmic membrane, and finally sorbitol is released from it, which is a procedure involving the help of cytoskeleton [19]. Its biological significance is related to cell volume regulation [13,83].

In this study, we showed AA caused a sharp rise in LDH leakage indicative of cell injury, while HP clearly inhibited it. In addition, BB, an inhibitor of AR, completely reversed the protection of HP. These results indicated that AR was involved in the protection produced by HP. AA furthermore caused the increase in cell volume, which would result in swelling and eventually lead cells to necrosis. By observing the change of cell volume at different time points, we found that HP not only delayed the appearance of RVD but also inhibited the increase of cell volume during 24 h of AA exposure. This suggested that cell volume regulation could be a potential mechanism in the protection exerted by HP against AA. We further demonstrated that HP significantly increased sorbitol levels, while the inhibitor of AR, BB, attenuated the increase in sorbitol content induced by HP. According to the above results, we hypothesized that sorbitol might be correlated with increased AR by HP. The fact that quinidine, a stronger inhibitor of sorbitol, reversed the protection afforded by HP indicates that sorbitol contributes to the protection of HP [95].

In summary, HIF-1, ROS and regulation of cell volume may mediate the protection of HP against anoxic or ischemic injury in CNS. Disclosure of the mechanisms of HP will contribute to the prevention and treatment of anoxic or ischemic diseases in brain.

5. Acknowledgments

These studies were supported by grants from the National Key Basic Research Program of China (2006CB504100 and 2012CB518200), the National Natural Science Foundation of China (81071066 and 81000856).

6. References

[1] Ara J, Fekete S, Frank M, Golden JA, Pleasure D, Valencia I.Hypoxic-preconditioning induces neuroprotection against hypoxia-ischemia in newborn piglet brain. Neurobiol Dis. 2011;43(2):473-485.

[2] Baranova O, Miranda LF, Pichiule P, Dragatsis I, Johnson RS, Chavez JC. Neuron-specific inactivation of the hypoxia inducible factor 1 alpha increases brain injury in a mouse model of transient focal cerebral ischemia. J Neurosci. 2007;27(23):6320-6332.

[3] Bellot G, Garcia-Medina R, Gounon P, Chiche J, Roux D, Pouysségur J, Mazure NM. Hypoxia-induced autophagy is mediated through hypoxia-inducible factor induction of BNIP3 and BNIP3L via their BH3 domains. Mol Cell Biol. 2009;29(10):2570-2581.

[4] Bergeron M, Ferriero DM, Vreman HJ, Stevenson DK, Sharp FR. Hypoxia-ischemia, but not hypoxia alone, induces the expression of heme oxygenase-1 (HSP32) in newborn rat brain. J Cereb Blood Flow Metab. 1997;17(6):647-658.

[5] Bergeron M, Gidday JM, Yu AY, Semenza GL, Ferriero DM, Sharp FR. Role of hypoxia-inducible factor-1 in hypoxia-induced ischemic tolerance in neonatal rat brain. Ann Neurol. 2000;48(3):285-296.

[6] Bernaudin M, Nedelec AS, Divoux D, MacKenzie ET, Petit E, Schumann-Bard P. Normobaric hypoxia induces tolerance to focal permanent cerebral ischemia in association with an increased expression of hypoxia-inducible factor-1 and its target genes, erythropoietin and VEGF, in the adult mouse brain. J Cereb Blood Flow Metab. 2002;22(4):393-403.

[7] Bernaudin M, Tang Y, Reilly M, Petit E, Sharp FR. Brain genomic response following hypoxia and re-oxygenation in the neonatal rat. Identification of genes that might contribute to hypoxia-induced ischemic tolerance. J Biol Chem. 2002;277(42):39728-8.

[8] Bickler PE, Fahlman CS, Gray J, McKleroy W. Inositol 1,4,5-triphosphate receptors and NAD(P)H mediate Ca2+ signaling required for hypoxic preconditioning of hippocampal neurons. Neuroscience. 2009;160(1):51-60.

[9] Boyd JM, Malstrom S, Subramanian T, Venkatesh LK, Schaeper U, Elangovan B, D'Sa-Eipper C, Chinnadurai G. Adenovirus E1B 19 kDa and Bcl-2 proteins interact with a common set of cellular proteins. Cell. 1994;79(6):341-351.

[10] Bruer U, Weih MK, Isaev NK, Meisel A, Ruscher K, Bergk A, Trendelenburg G, Wiegand F, Victorov IV, Dirnagl U. Induction of tolerance in rat cortical neurons: hypoxic preconditioning. FEBS Lett. 1997;414(1):117-121.

[11] Brunelle JK, Bell EL, Quesada NM, Vercauteren K, Tiranti V, Zeviani M, Scarpulla RC, Chandel NS. Oxygen sensing requires mitochondrial ROS but not oxidative phosphorylation. Cell Metab. 2005;1(6):409-414.

[12] Bu X, Huang P, Qi Z, Zhang N, Han S, Fang L, Li J. Cell type-specific activation of p38 MAPK in the brain regions of hypoxic preconditioned mice. Neurochem Int. 2007;51(8):459-466.

[13] Burg MB. Molecular basis of osmotic regulation. Am J Physiol. 1995;268(6 Pt 2):F983-996. Review.

[14] Centeno JM, Orti M, Salom JB, Sick TJ, Pérez-Pinzón MA. Nitric oxide is involved in anoxic preconditioning neuroprotection in rat hippocampal slices. Brain Res. 1999;836(1-2):62-69.

[15] Chang S, Jiang X, Zhao C, Lee C, Ferriero DM. Exogenous low dose hydrogen peroxide increases hypoxia-inducible factor-1alpha protein expression and induces preconditioning protection against ischemia in primary cortical neurons. Neurosci Lett. 2008;441(1):134-138.

[16] Chin BY, Jiang G, Wegiel B, Wang HJ, Macdonald T, Zhang XC, Gallo D, Cszimadia E, Bach FH, Lee PJ, Otterbein LE. Hypoxia-inducible factor 1alpha stabilization by carbon monoxide results in cytoprotective preconditioning. Proc Natl Acad Sci U S A. 2007;104(12):5109-5114.

[17] Choi KO, Lee T, Lee N, Kim JH, Yang EG, Yoon JM, Kim JH, Lee TG, Park H. Inhibition of the catalytic activity of hypoxia-inducible factor-1alpha-prolyl-hydroxylase 2 by a MYND-type zinc finger. Mol Pharmacol. 2005;68(6):1803-1809.

[18] Churilova AV, Rybnikova EA, Glushchenko TS, Tyulkova EI, Samoilov MO.Effects of moderate hypobaric hypoxic preconditioning on the expression of the transcription factors pCREB and NF-kappaB in the rat hippocampus before and after severe hypoxia. Neurosci Behav Physiol. 2010;40(8):852-857.

[19] Czekay RP, Kinne-Saffran E, Kinne RK. Membrane traffic and sorbitol release during osmo- and volume regulation in isolated rat renal inner medullary collecting duct cells. Eur J Cell Biol. 1994;63(1):20-31.

[20] Digicaylioglu M, Lipton SA. Erythropoietin-mediated neuroprotection involves cross-talk between Jak2 and NF-kappaB signalling cascades. Nature. 2001;412(6847):641-647.

[21] Doedens A, Johnson RS. Transgenic models to understand hypoxia-inducible factor function. Methods Enzymol. 2007;435:87-105.

[22] Emerling BM, Platanias LC, Black E, Nebreda AR, Davis RJ, Chandel NS. Mitochondrial reactive oxygen species activation of p38 mitogen-activated protein kinase is required for hypoxia signaling. Mol Cell Biol. 2005;25(12):4853-4862.

[23] Essick EE, Sam F. Oxidative stress and autophagy in cardiac disease, neurological disorders, aging and cancer. Oxid Med Cell Longev. 2010;3(3):168-177. Review.

[24] Favaro E, Ramachandran A, McCormick R, Gee H, Blancher C, Crosby M, Devlin C, Blick C, Buffa F, Li JL, Vojnovic B, Pires das Neves R, Glazer P, Iborra F, Ivan M, Ragoussis J, Harris AL. MicroRNA-210 regulates mitochondrial free radical response to hypoxia and krebs cycle in cancer cells by targeting iron sulfur cluster protein ISCU. PLoS One. 2010;5(4):e10345.

[25] Garnier P, Demougeot C, Bertrand N, Prigent-Tessier A, Marie C, Beley A. Stress response to hypoxia in gerbil brain: HO-1 and Mn SOD expression and glial activation. Brain Res. 2001;893(1-2):301-309.

[26] Garty H, Furlong TJ, Ellis DE, Spring KR. Sorbitol permease: an apical membrane transporter in cultured renal papillary epithelial cells. Am J Physiol. 1991;260(5 Pt 2):F650-656.

[27] Gidday JM, Fitzgibbons JC, Shah AR, Park TS. Neuroprotection from ischemic brain injury by hypoxic preconditioning in the neonatal rat. Neurosci Lett. 1994;168(1-2):221-224.
[28] Gorgias N, Maidatsi P, Tsolaki M, Alvanou A, Kiriazis G, Kaidoglou K, Giala M. Hypoxic pretreatment protects against neuronal damage of the rat hippocampus induced by severe hypoxia. Brain Res. 1996 ;714(1-2):215-225.
[29] Grimm C, Hermann DM, Bogdanova A, Hotop S, Kilic U, Wenzel A, Kilic E, Gassmann M. Neuroprotection by hypoxic preconditioning: HIF-1 and erythropoietin protect from retinal degeneration. Semin Cell Dev Biol. 2005;16(4-5):531-538.
[30] Grimm C, Wenzel A, Acar N, Keller S, Seeliger M, Gassmann M. Hypoxic preconditioning and erythropoietin protect retinal neurons from degeneration. Adv Exp Med Biol. 2006;588:119-131. Review.
[31] Hamacher-Brady A, Brady NR, Logue SE, Sayen MR, Jinno M, Kirshenbaum LA, Gottlieb RA, Gustafsson AB. Response to myocardial ischemia/reperfusion injury involves Bnip3 and autophagy. Cell Death Differ. 2007;14(1):146-157.
[32] Hamanaka RB, Chandel NS. Mitochondrial reactive oxygen species regulate hypoxic signaling. Curr Opin Cell Biol. 2009;21(6):894-899. Review.
[33] Harten SK, Ashcroft M, Maxwell PH. Prolyl hydroxylase domain inhibitors: a route to HIF activation and neuroprotection. Antioxid Redox Signal. 2010;12(4):459-480.
[34] Helton R, Cui J, Scheel JR, Ellison JA, Ames C, Gibson C, Blouw B, Ouyang L, Dragatsis I, Zeitlin S, Johnson RS, Lipton SA, Barlow C. Brain-specific knock-out of hypoxia-inducible factor-1alpha reduces rather than increases hypoxic-ischemic damage. J Neurosci. 2005;25(16):4099-4107.
[35] Hirota K, Semenza GL. Regulation of hypoxia-inducible factor 1 by prolyl and asparaginyl hydroxylases. Biochem Biophys Res Commun. 2005;338(1):610-616. Review.
[36] Huang P, Qi Z, Bu X, Zhang N, Han S, Fang L, Li J. Neuron-specific phosphorylation of mitogen- and stress-activated protein kinase-1 involved in cerebral hypoxic preconditioning of mice. J Neurosci Res. 2007;85(6):1279-1287.
[37] Hwang YC, Shaw S, Kaneko M, Redd H, Marrero MB, Ramasamy R. Aldose reductase pathway mediates JAK-STAT signaling: a novel axis in myocardial ischemic injury. FASEB J. 2005;19(7):795-797.
[38] Jian Z, Li K, Liu L, Zhang Y, Zhou Z, Li C, Gao T. Heme Oxygenase-1 Protects Human Melanocytes from H(2)O(2)-Induced Oxidative Stress via the Nrf2-ARE Pathway. J Invest Dermatol. 2011;131(7):1420-1427.
[39] Jiang J, Yang W, Huang P, Bu X, Zhang N, Li J. Increased phosphorylation of Ets-like transcription factor-1 in neurons of hypoxic preconditioned mice. Neurochem Res. 2009;34(8):1443-1450.
[40] Jones NM, Bergeron M. Hypoxic preconditioning induces changes in HIF-1 target genes in neonatal rat brain. J Cereb Blood Flow Metab. 2001;21:1105–1114.
[41] Jones NM, Lee EM, Brown TG, Jarrott B, Beart PM. Hypoxic preconditioning produces differential expression of hypoxia-inducible factor-1alpha (HIF-1alpha) and its regulatory enzyme HIF prolyl hydroxylase 2 in neonatal rat brain. Neurosci Lett. 2006;404(1-2):72-77.
[42] Karhausen J, Kong T, Narravula S, Colgan SP. Induction of the von Hippel-Lindau tumor suppressor gene by late hypoxia limits HIF-1 expression. J Cell Biochem. 2005;95(6):1264-1275.

[43] Kim HW, Haider HK, Jiang S, Ashraf M. Ischemic preconditioning augments survival of stem cells via miR-210 expression by targeting caspase-8-associated protein 2. J Biol Chem. 2009;284(48):33161-33168.
[44] Kim I, Rodriguez-Enriquez S, Lemasters JJ. Selective degradation of mitochondria by mitophagy. Arch Biochem Biophys. 2007;462(2):245-253. Review.
[45] Kitagawa K, Matsumoto M, Tagaya M, Hata R, Ueda H, Niinobe M, Handa N, Fukunaga R, Kimura K, Mikoshiba K. 'Ischemic tolerance' phenomenon found in the brain. Brain Res. 1990,528(1):21–24.
[46] Lai IR, Chang KJ, Tsai HW, Chen CF. Pharmacological preconditioning with simvastatin protects liver from ischemia-reperfusion injury by heme oxygenase-1 induction. Transplantation. 2008;85(5):732-738.
[47] Lai IR, Ma MC, Chen CF, Chang KJ. The protective role of heme oxygenase-1 on the liver after hypoxic preconditioning in rats. Transplantation. 2004;77(7):1004-1008.
[48] Lambrechts D, Storkebaum E, Carmeliet P. VEGF: necessary to prevent motoneuron degeneration, sufficient to treat ALS? Trends Mol Med. 2004;10(6):275-282.
[49] Lando D, Peet DJ, Gorman JJ, Whelan DA, Whitelaw ML, Bruick RK. FIH-1 is an asparaginyl hydroxylase enzyme that regulates the transcriptional activity of hypoxia-inducible factor. Genes Dev. 2002;16(12):1466-1471.
[50] Leconte C, Tixier E, Freret T, Toutain J, Saulnier R, Boulouard M, Roussel S, Schumann-Bard P, Bernaudin M. Delayed hypoxic postconditioning protects against cerebral ischemia in the mouse. Stroke. 2009;40(10):3349-3355.
[51] Li QF, Zhu YS, Jiang H. Isoflurane preconditioning activates HIF-1alpha, iNOS and Erk1/2 and protects against oxygen-glucose deprivation neuronal injury. Brain Res. 2008;1245:26-35.
[52] Li RC, Guo SZ, Lee SK, Gozal D. Neuroglobin protects neurons against oxidative stress in global ischemia. J Cereb Blood Flow Metab. 2010;30(11):1874-1882.
[53] Liu J, Ginis I, Spatz M, Hallenbeck JM. Hypoxic preconditioning protects cultured neurons against hypoxic stress via TNF-alpha and ceramide. Am J Physiol Cell Physiol. 2000;278(1):C144-153.
[54] Liu J, Narasimhan P, Yu F, Chan PH. Neuroprotection by hypoxic preconditioning involves oxidative stress-mediated expression of hypoxia-inducible factor and erythropoietin. Stroke. 2005;36(6):1264-1269.
[55] Luo T, Zhang H, Zhang WW, Huang JT, Song EL, Chen SG, He F, Xu J, Wang HQ. Neuroprotective effect of Jatrorrhizine on hydrogen peroxide-induced cell injury and its potential mechanisms in PC12 cells. Neurosci Lett. 2011 May 13. [Epub ahead of print]
[56] Lushnikova I, Orlovsky M, Dosenko V, Maistrenko A, Skibo G. Brief anoxia preconditioning and HIF prolyl-hydroxylase inhibition enhances neuronal resistance in organotypic hippocampal slices on model of ischemic damage. Brain Res. 2011;1386:175-183.
[57] MacFarlane PM, Wilkerson JE, Lovett-Barr MR, Mitchell GS. Reactive oxygen species and respiratory plasticity following intermittent hypoxia. Respir Physiol Neurobiol. 2008;164(1-2):263-271.
[58] Macmanus CF, Campbell EL, Keely S, Burgess A, Kominsky DJ, Colgan SP. Anti-inflammatory actions of adrenomedullin through fine tuning of HIF stabilization. FASEB J. 2011;25(6):1856-1864.

[59] Mahon PC, Hirota K, Semenza GL. FIH-1: a novel protein that interacts with HIF-1alpha and VHL to mediate repression of HIF-1 transcriptional activity. Genes Dev. 2001;15(20):2675-2686.
[60] Mazure NM, Pouysségur J. Hypoxia-induced autophagy: cell death or cell survival? Curr Opin Cell Biol. 2010;22(2):177-180.
[61] Miller BA, Perez RS, Shah AR, Gonzales ER, Park TS, Gidday JM. Cerebral protection by hypoxic preconditioning in a murine model of focal ischemia-reperfusion. Neuroreport. 2001;12(8):1663-1669.
[62] Muñoz A, Nakazaki M, Goodman JC, Barrios R, Onetti CG, Bryan J, Aguilar-Bryan L. Ischemic preconditioning in the hippocampus of a knockout mouse lacking SUR1-based K(ATP) channels. Stroke. 2003;34(1):164-170.
[63] Nam SY, Sabapathy K.p53 promotes cellular survival in a context-dependent manner by directly inducing the expression of haeme-oxygenase-1. Oncogene. 2011 May 9. [Epub ahead of print]
[64] Nishijima K, Ng YS, Zhong L, Bradley J, Schubert W, Jo N, Akita J, Samuelsson SJ, Robinson GS, Adamis AP, Shima DT. Vascular endothelial growth factor-A is a survival factor for retinal neurons and a critical neuroprotectant during the adaptive response to ischemic injury. Am J Pathol. 2007;171(1):53-67.
[65] Obrenovitch TP. Molecular physiology of preconditioning-induced brain tolerance to ischemia. Physiol Rev. 2008;88(1):211-247. Review.
[66] Okinaga S, Shibahara S. Identification of a nuclear protein that constitutively recognizes the sequence containing a heat-shock element. Its binding properties and possible function modulating heat-shock induction of the rat heme oxygenase gene. Eur J Biochem. 1993; 212: 167-175.
[67] Omata N, Murata T, Takamatsu S, Maruoka N, Wada Y, Yonekura Y, Fujibayashi Y. Hypoxic tolerance induction in rat brain slices following hypoxic preconditioning due to expression of neuroprotective proteins as revealed by dynamic changes in glucose metabolism. Neurosci Lett. 2002;329(2):205-208.
[68] Oosthuyse B, Moons L, Storkebaum E, Beck H, Nuyens D, Brusselmans K, Van Dorpe J, Hellings P, Gorselink M, Heymans S, Theilmeier G, Dewerchin M, Laudenbach V, Vermylen P, Raat H, Acker T, Vleminckx V, Van Den Bosch L, Cashman N, Fujisawa H, Drost MR, Sciot R, Bruyninckx F, Hicklin DJ, Ince C, Gressens P, Lupu F, Plate KH, Robberecht W, Herbert JM, Collen D, Carmeliet P. Deletion of the hypoxia-response element in the vascular endothelial growth factor promoter causes motor neuron degeneration. Nat Genet. 2001;28(2):131-138.
[69] Patten DA, Lafleur VN, Robitaille GA, Chan DA, Giaccia AJ, Richard DE. Hypoxia-inducible factor-1 activation in nonhypoxic conditions: the essential role of mitochondrial-derived reactive oxygen species. Mol Biol Cell. 2010;21(18):3247-3257.
[70] Peet D, Linke S. Regulation of HIF: asparaginyl hydroxylation. Novartis Found Symp. 2006;272:37-49; discussion 49-53, 131-40. Review.
[71] Prass K, Scharff A, Ruscher K, Löwl D, Muselmann C, Victorov I, Kapinya K, Dirnagl U, Meisel A. Hypoxia-induced stroke tolerance in the mouse is mediated by erythropoietin. Stroke. 2003;34(8):1981-1986.
[72] Ryu JH, Li SH, Park HS, Park JW, Lee B, Chun YS. Hypoxia-inducible factor α subunit stabilization by NEDD8 conjugation is reactive oxygen species-dependent. J Biol Chem. 2011;286(9):6963-6970.

[73] Safran M, Kim WY, O'Connell F, Flippin L, Günzler V, Horner JW, Depinho RA, Kaelin WG Jr. Mouse model for noninvasive imaging of HIF prolyl hydroxylase activity: assessment of an oral agent that stimulates erythropoietin production. Proc Natl Acad Sci U S A. 2006;103(1):105-110.
[74] Samoilov MO, Lazarevich EV, Semenov DG, Mokrushin AA, Tyul'kova EI, Romanovskii DY, Milyakova EA, Dudkin KN. The adaptive effects of hypoxic preconditioning of brain neurons. Neurosci Behav Physiol. 2003;33(1):1-11. Review.
[75] Scherz-Shouval R, Elazar Z. Regulation of autophagy by ROS: physiology and pathology. Trends Biochem Sci. 2011;36(1):30-38. Review.
[76] Schüttert JB, Fiedler GM, Grupp C, Blaschke S, Grunewald RW. Sorbitol transport in rat renal inner medullary interstitial cells. Kidney Int. 2002;61(4):1407-1415.
[77] Semenza GL. HIF-1: mediator of physiological and pathological responses to hypoxia. J Appl Physiol. 2000;88:1474–1480.
[78] Semenza GL. Hypoxia-inducible factor 1: oxygen homeostasis and disease pathophysiology. Trends Mol Med. 2001;7(8):345-350.
[79] Semenza GL. Hypoxia-inducible factor 1: Regulator of mitochondrial metabolism and mediator of ischemic preconditioning. Biochim Biophys Acta. 2011;1813(7):1263-1268.
[80] Sharp FR, Bergeron M, Bernaudin M. Hypoxia-inducible factor in brain. Adv Exp Med Biol. 2001;502:273-291.
[81] Sharp FR, Bernaudin M. HIF1 and oxygen sensing in the brain. Nat Rev Neurosci 2004;5:437-448.
[82] Sharp FR, Ran R, Lu A, Tang Y, Strauss KI, Glass T, Ardizzone T, Bernaudin M. Hypoxic preconditioning protects against ischemic brain injury. NeuroRx. 2004;1(1):26-35.
[83] Siebens AW, Spring KR. A novel sorbitol transport mechanism in cultured renal papillary epithelial cells. Am J Physiol. 1989;257(6 Pt 2):F937-946.
[84] Simon MC. Mitochondrial reactive oxygen species are required for hypoxic HIF alpha stabilization. Adv Exp Med Biol. 2006;588:165-170. Review.
[85] Sowter HM, Ratcliffe PJ, Watson P, Greenberg AH, Harris AL. HIF-1-dependent regulation of hypoxic induction of the cell death factors BNIP3 and NIX in human tumors. Cancer Res. 2001;61(18):6669-6673.
[86] Stolze IP, Mole DR, Ratcliffe PJ. Regulation of HIF: prolyl hydroxylases. Novartis Found Symp. 2006;272:15-25; discussion 25-36. Review.
[87] Taie S, Ono J, Iwanaga Y, Tomita S, Asaga T, Chujo K, Ueki M. Hypoxia-inducible factor-1 alpha has a key role in hypoxic preconditioning. J Clin Neurosci. 2009;16(8):1056-1060.
[88] Tang Y, Pacary E, Fréret T, Divoux D, Petit E, Schumann-Bard P, Bernaudin M. Effect of hypoxic preconditioning on brain genomic response before and following ischemia in the adult mouse: identification of potential neuroprotective candidates for stroke. Neurobiol Dis. 2006;21(1):18-28.
[89] Theus MH, Wei L, Cui L, Francis K, Hu X, Keogh C, Yu SP. In vitro hypoxic preconditioning of embryonic stem cells as a strategy of promoting cell survival and functional benefits after transplantation into the ischemic rat brain. Exp Neurol. 2008;210(2):656-670.
[90] Tzeng YW, Lee LY, Chao PL, Lee HC, Wu RT, Lin AM. Role of autophagy in protection afforded by hypoxic preconditioning against MPP+-induced neurotoxicity in SH-SY5Y cells. Free Radic Biol Med. 2010;49(5):839-846.

[91] Wang GL, Semenza GL. General involvement of hypoxia-inducible factor 1 in transcriptional response to hypoxia. Proc Natl Acad Sci U S A. 1993;90(9):4304-4308.

[92] Wick A, Wick W, Waltenberger J, Weller M, Dichgans J, Schulz JB. Neuroprotection by hypoxic preconditioning requires sequential activation of vascular endothelial growth factor receptor and Akt. J Neurosci. 2002;22(15):6401-6407.

[93] Wu LY, Ding AS, Zhao T, Ma ZM, Wang FZ, Fan M.Involvement of increased stability of mitochondrial membrane potential and overexpression of Bcl-2 in enhanced anoxic tolerance induced by hypoxic preconditioning in cultured hypothalamic neurons. Brain Res. 2004;999(2):149-154.

[94] Wu LY, Ding AS, Zhao T, Ma ZM, Wang FZ, Fan M.Underlying mechanism of hypoxic preconditioning decreasing apoptosis induced by anoxia in cultured hippocampal neurons. Neurosignals. 2005;14(3):109-116.

[95] Wu LY, Ma ZM, Fan XL, Zhao T, Liu ZH, Huang X, Li MM, Xiong L, Zhang K, Zhu LL, Fan M. The anti-necrosis role of hypoxic preconditioning after acute anoxia is mediated by aldose reductase and sorbitol pathway in PC12 cells. Cell Stress Chaperones. 2010 ;15(4):387-394.

[96] Yu S, Zhao T, Guo M, Fang H, Ma J, Ding A, Wang F, Chan P, Fan M. Hypoxic preconditioning up-regulates glucose transport activity and glucose transporter (GLUT1 and GLUT3) gene expression after acute anoxic exposure in the cultured rat hippocampal neurons and astrocytes. Brain Res. 2008;1211:22-29.

[97] Zhang H, Bosch-Marce M, Shimoda LA, Tan YS, Baek JH, Wesley JB, Gonzalez FJ, Semenza GL. Mitochondrial autophagy is an HIF-1-dependent adaptive metabolic response to hypoxia. J. J Biol Chem. 2008;283(16):10892-10903.

[98] Zheng Y, Liu Y, Ge J, Wang X, Liu L, Bu Z, Liu P. Resveratrol protects human lens epithelial cells against H2O2-induced oxidative stress by increasing catalase, SOD-1, and HO-1 expression. Mol Vis. 2010;16:1467-1474.

[99] Zhu Y, Zhang Y, Ojwang BA, Brantley MA Jr, Gidday JM. Long-term tolerance to retinal ischemia by repetitive hypoxic preconditioning: role of HIF-1alpha and heme oxygenase-1. Invest Ophthalmol Vis Sci. 2007;48(4):1735-1743.

[100] Valko M, Leibfritz D, Moncol J, Cronin MT, Mazur M, Telser J. Free radicals and antioxidants in normal physiological functions and human disease. Int J Biochem Cell Biol. 2007;39:44–84.

[101] Desikan R, Hancock, Neill SJ. Oxidative stress signaling. Topics in Current Genetics, 2004; 4:121-149.

[102] Semenza GL. Autophagy. Mitochondrial autophagy: life and breath of the cell. 2008; 4(4):534-536.

[103] Kaushik S and Cuervo AM. Autophagy as a cell-repair mechanism: activation of chaperone-mediated autophagy during oxidative stress. Mol Aspects Med. 2006;27, 444-454.

[104] Fukuda R, Zhang H, Kim JW, Shimoda L, Dang CV, Semenza GL. HIF-1 regulates cytochrome oxidase subunits to optimize efficiency of respiration in hypoxic cells. Cell. 2007;129(1):111-122.

[105] Kim JW, Tchernyshyov I, Semenza GL, Dang CV. HIF-1-mediated expression of pyruvate dehydrogenase kinase: a metabolic switch required for cellular adaptation to hypoxia. Cell Metab. 2006;3(3):177-185.

Part 3

Ecological Systems

5

Increase in Anoxia in Lake Victoria and Its Effects on the Fishery

Murithi Njiru[1], Chrisphine Nyamweya[2], John Gichuki[2], Rose Mugidde[3], Oliva Mkumbo[4] and Frans Witte[5]
[1]Department of Fisheries and Aquatic Sciences, Moi University, Eldoret,
[2]Kenya Marine and Fisheries Research Institute, Kisumu, Kenya
[3]COWI, Uganda Ltd, Plot No. 3, Portal Avenue, Kampala,
[4]Lake Victoria Fisheries Organisation (LVFO), Jinja,
[5]Institute of Biology Leiden, Leiden University, Leiden,
[1,2]Kenya
[3,4]Uganda
[5]Netherlands

1. Introduction

Lake Victoria is the second largest freshwater lake in the world and the largest in the tropics (Crul, 1995). It is found in East Africa, within 0°20'N to 3°00'S and 31°39'E to 34°53'E at an altitude of 1134 m. It has a surface area of about 68,800 km^2 and a maximum depth of about 70 m. The lake is shared by Tanzania (51%), Uganda (43%), Kenya (6%), with a drainage basin of about 195 000 km^2 which includes the neighbouring states of Rwanda and Burundi. Lake Victoria originated less than 1 million years ago and may have been dry for several millennia until about 14,000 years ago (Stager & Johnson, 2007). The lake and its numerous satellite lakes were originally dominated by a rich fish fauna, comprising several hundred species of cichlids (Kaufman et al., 1997).

Until the 1960s, the tilapiine species *Oreochromis esculentus* (Graham) and *Oreochromis variabilis* (Boulenger) were the most important commercial species. Other important species included *Protopterus aethiopicus* Heckel, *Bagrus docmak* (Forsskål), *Clarias gariepinus* (Burchell), various *Barbus* species, mormyrids and *Schilbe intermedius* Rüppell. *Labeo victorianus* Boulenger formed the most important commercial species in the affluent rivers of the lake basin. Haplochromine cichlids and a native pelagic cyprinid, *Rastrineobola argentea* (Pellegrin), were abundant, but were not originally exploited on a large scale because of their small size (Kudhogania & Cordone, 1974; Ogutu-Ohwayo, 1990). In the late 1950s and early 1960s, the predatory Nile perch *Lates niloticus* (L) and four tilapiine species, Nile tilapia, *Oreochromis niloticus* (L), *Oreochromis leucostictus* (Trewavas), *Tilapia zillii* (Gervais) and *Tilapia rendalii* Boulenger, were introduced into the Lake Victoria basin to increase depleted commercial fisheries (Ogutu-Ohwayo, 1990, 1994). Thereafter, the fish community of lake with over 50% of native fish species (> 99% endemic haplochromines cichlids) disappeared (Ogutu-Ohwayo, 1990; Witte et al., 1992a). Decline in haplochromines was attributed mainly to the predatory Nile perch (Ogutu-Ohwayo,

1990). Nile tilapia which was better competitor for resources hybridized and displaced the indigenous oreochormine group. The once multispecies fishery was converted to three species, viz Nile perch, Nile tilapia and *R. argentea,* changing drastically the food web. However, the introduction of Nile perch transformed the fisheries from a locally based artisanal fishery to an international capital investment industry. The fisheries produces an annual income of approximately $US 600 million, providing employment opportunities for over 3 million people (Njiru et al., 2005).

For the past few decades, Lake Victoria basin has undergone environmental degradation leading to increased nutrients inputs into the lake leading to eutrophication, (Hecky et al., 1994). Increased eutrophication has led to increased algal blooms and deoxygenation in the deeper waters (Talling, 1966, Hecky et al., 1994). Hypoxia which occurred only in deep waters in the 1960s (Talling, 1966), is now more frequent and is spreading into shallow waters (Hecky et al., 1994, Mugidde et al., 2005). Deoxygenation, intense fishing and predation by Nile perch contributed to the demise of many native species (Ogutu-Ohwayo 1990, 1994; Kaufman & Ochumba, 1993). This chapter explores the impact of hypoxia on the fishery of Lake Victoria.

2. Fish and dissolved oxygen

Oxygen is necessary to sustain life of fishes dependent on aerobic respiration (Diaz & Breitburg, 2009). Availability of dissolved oxygen (DO) is one abiotic factor that can limit habitat quality, distribution, growth, reproduction, and survival of fishes (Kramer, 1987). Though all fishes require oxygen for survival, the physical properties of water (high viscosity, low oxygen content at saturation) makes its uptake challenging for fishes even at high DO levels (Kramer, 1987). When the supply of oxygen is cut off or consumption exceeds resupply, DO concentration can decline below levels required by most animal life. The condition of low DO is known as hypoxia while water devoid of oxygen is referred to as anoxic. Hypoxia and anoxia differ quantitatively in the availability of oxygen, as well as qualitatively in the presence of toxic compounds such as hydrogen sulfide. Hypoxia can be described in several forms (Farrell & Richards, 2009; Diaz & Breitburg, 2009). Aquatic hypoxia can be defined as dissolved oxygen concentrations below 5-6 mg l^{-1} in freshwater, 2–3 mg l^{-1} in marine and estuarine environments.

Hypoxia is a natural component of many freshwater habitats. Hypoxic conditions can be due to natural causes such as algal respiration, seasonal flooding, stratification, and anthropogenic causes. Low dissolved oxygen environments vary in temporal frequency, seasonality, and persistence (Farrell & Richards, 2009; Diaz & Breitburg, 2009; Wetzel, 2001). Hypoxia occurs naturally in habitats characterized by low mixing or light limited, heavily vegetated swamps and backwaters that circulate poorly, stratify, and have large loads of terrestrial organic matter (Kramer, 1987; Chapman et al., 1999). Levels of hypoxia are mainly determined by primary productivity, depth, and temperature of the aquatic body (Wetzel, 2001).

Hypoxia is physiologically stressful for fish, shellfish, and invertebrates with prolonged exposure to anoxia being fatal to most aquatic fauna (Diaz & Breitburg, 2009; Chapman & McKenzie, 2009). Hypoxia exposure can prompt both lethal and sublethal effects in fishes, leading to reduced feeding, reproductive, growth, metabolism, and slower reaction time. These effects vary across fish species (Chapman & McKenzie, 2009), but also depend on the frequency, intensity, and duration of the hypoxic events (Diaz & Breitburg, 2009). Fishes can compensate for hypoxic conditions by decreasing their need for oxygen, increasing the amount of oxygen available, or combining both strategies (Reardon & Chapman, 2010).

3. Hypoxia in Lake Victoria

Hypoxic (and anoxic) environments in aquatic systems have existed through geological time (Chapman & McKenzie, 2009). However, environmental degradation and global climate changes is increasing the occurrence of hypoxia. Increase in inputs from municipal wastes and agricultural runoffs have accelerated eutrophication of water bodies. This has altered the balance of aquatic ecosystem productivity and metabolic activities of bacteria leading to hypoxic conditions. In, Lake Victoria, hypoxia was first recorded in the late 1920s (Worthington, 1930). Fossil evidence indicated that the hypoxia in the lake could be attributed to continued increase in nutrients inputs, and altered food-web following the boom of predaceous Nile perch in the 1980s (Verschuren et al., 2002). Palaeolimnological data showed an increase in phytoplankton production from 1930s, paralleling human-population growth and agricultural activity in the lake basin (Verschuren et al., 2002). Using evidence from changes in the *Procladius/ Chironomus* ratio of fossil midge faunas, Verschuren et al., (2002) reported that the deep-water oxygen regime of Lake Victoria started to deteriorate in the early 1960s. Seasonally persistent deep-water anoxia appeared to have reached its current level by the late 1970s. These fossil data agree with historical dissolved oxygen data from the 1920s, 1960s and 1990s (Worthington, 1930; Hecky et al., 1994). The deep-water anoxia first observed in 1960–1961 (Talling, 1966), probably represented the earliest stage of eutrophication-induced deepwater oxygen loss in Lake Victoria (Figure 1a). Currently hypoxia has spread to shallow areas of the lake (Figure 1b). Studies have found 3 mgl^{-1} to be the critical DO concentration below which most fish in Lake Victoria will not survive (Kaufman, 1992, Wanink et al., 2001).

Calamari et al., (1995) noted that phosphorus was the limiting factor for primary production in Lake Victoria. Discharge of phosphorus into the lake is possibly the main factor leading to eutrophication, with its concentration having increased by a factor of 2 to 3 (Mugidde et al., 2005). Phosphorous which disrupts normal biogeochemical cycles is mainly from human activities like; deforestation, agricultural activities and municipal wastes. High nutrient concentration support elevated primary production and algal biomass has risen by up to a factor of 8 (Mugidde et al., 2005; Lung'ayia et al., 2000). As more organic matter is produced more oxygen is needed to remineralize the organics, primarily through the microbial loop. The lake ecosystem becomes overloaded and DO declines leading to hypoxia and anoxic conditions. Earlier on, the balance of lake productivity and its water quality was maintained by its faunal composition (Goldschmidt et al., 1993). However, anthropogenic activities and introduction of exotic species changed this balance. The lake could no longer maintain its integrity; it changed from a mesotrophic system in the 1920s, dominated by diatoms, to a eutrophic system dominated by blue-green algae (Hecky et al., 1994; Lung'ayia et al., 2000).

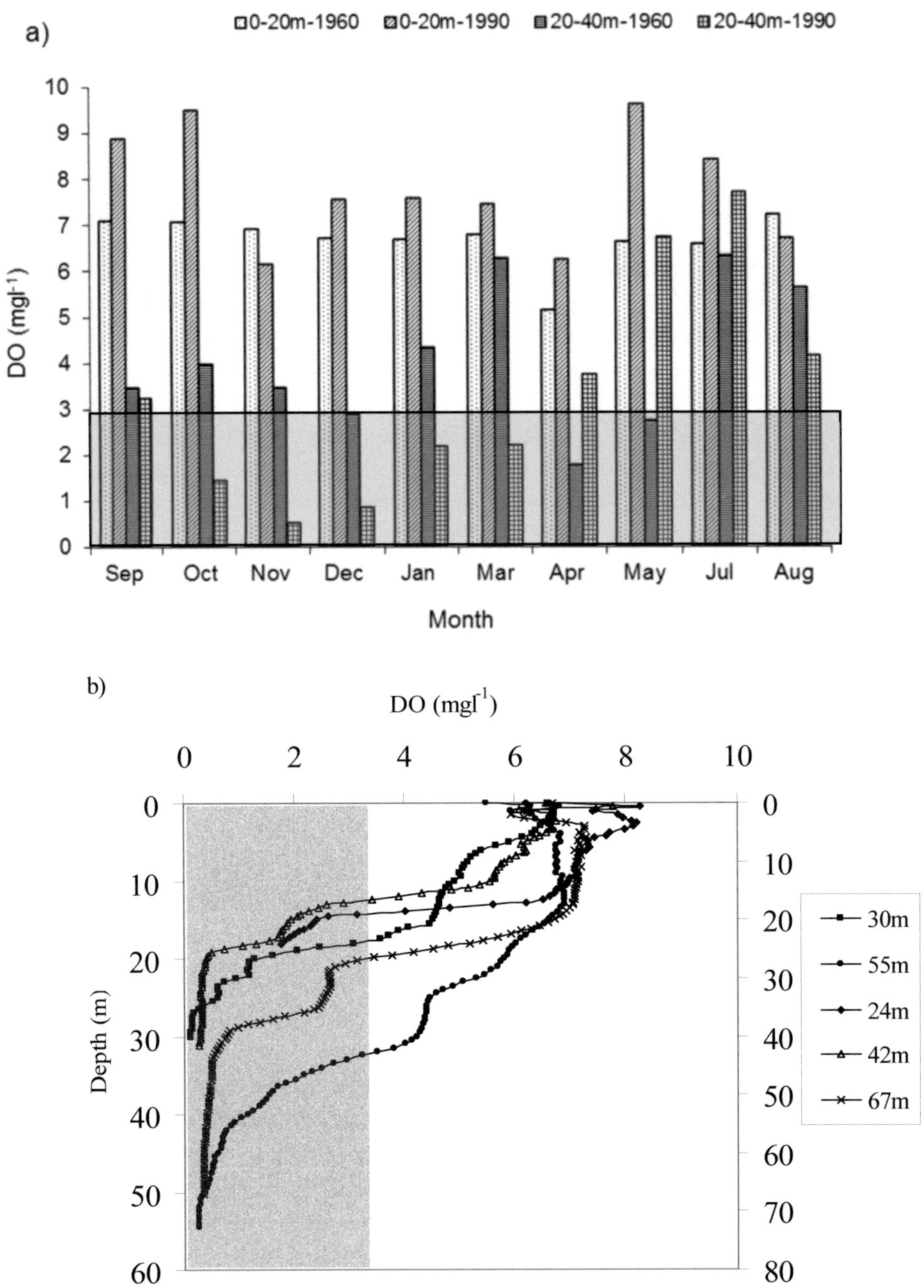

Fig. 1. Oxygen conditions in Lake Victoria, a) monthly oxygen levels with depth during 1960-61 (Talling, 1966) and 1990-1991 (Hecky et al., 1994), b) Oxygen profiles in February 2002 during stratification. The vertical grey bar represent oxygen concentration (<3 mgl^{-1}) considered critical for fish in the lake. Data Source, Kenya Marine and Fisheries Research Institute (KMFRI).

4. Stratification and hypoxia

Depending on temperature and salinity, water contains 20-40 times less oxygen by volume and diffuses about ten thousand times more slowly through water than air (Graham, 1990). The relative low solubility and diffusion of oxygen in water combined with density stratification and decomposition lead to the development of hypoxia. Due to low solubility, small changes in amount of DO in water such as from microbial and macrofaunal respiration leads to large differences in oxygen saturation in water. For example, 9.1 mg O_2 l^{-1} will dissolve in a liter of freshwater at 20°C; at this temperature a 1 mg O_2 l^{-1} drop in oxygen is equivalent to an 11% decline in oxygen saturation (Graham, 1990). Stratification of the water column isolates the bottom water from exchange with oxygen-rich surface water and the atmosphere, while decomposition of organic matter in the isolated bottom water consumes dissolved oxygen. For lakes, factors affecting vertical water mixing such as wind and temperature can lower DO in bottom waters to anoxic levels.

In Lake Victoria, thermal stratification, leading to hypoxia, was observed in the late 1920s (Worthington, 1930). Hypoxia conditions were restricted to the deeper waters (>60 m) and for shorter periods during the rainy season (Talling, 1966). Lake Victoria has three phases of thermal stratification; moderate stratification occurs between September and December, stable stratification in January to March/April, deep strong mixing in June to July (Talling, 1966; Mugidde et al., 2005). Global climatic change has affected this stratification, with the lake now warmer and thermal stratification more stable than in 1960s (Hecky et al., 1994). The overall minimum water temperature during the mixing period in June to July is 0.5 °C warmer than in the 1960s (Hecky et al., 1994). Increase in deep-water temperatures, increases thermal stratification stability (Hecky et al., 1994). Thermal stability makes the lake less able to mix effectively and promotes low oxygen conditions in deep waters during stratification period between September and April (Tables 1, 2; Mugidde et al., 2005).

Month	Temperature (°C)		Dissolved oxygen (mgl^{-1})		Mixing depth
	surface	bottom	surface	bottom	(m)
January	27.4	26.4	8.3	**2.1**	12.0
March	28.1	26.9	9.7	**4.4**	8.0
July	25.3	24.3	10.6	6.2	20.0
August	25.9	25.3	6.6	0.04	10.0
October	25.6	25.4	5.8	**1.3**	6.0
November	27.7	25.8	9.8	**0.5**	6.0
December	26.8	23.9	7.0	**0.6**	8.0

Table 1. Temperature and dissolved oxygen from an inshore shallow station (Napoleon Gulf, Uganda) in Lake Victoria. Boldface represents bottom DO levels during stratification period. Source: Mugidde et al. (2005).

Elevated water temperature accelerates algal productivity, chemical reactions and microbial processes such as denitrification-nitrification affecting nitrogen cycling and availability, resulting in a build up of regenerated nutrients in bottom waters. Resulting low oxygen conditions allow return of phosphorus from bottom to surface, but not nitrogen (Hecky et al., 1994, Mugidde et al., 2005). Continuous N loss due to denitrification during persistent stratification leads to high N demand and favours growth of blue-greens including N-fixers

Month	Temperature (°C) surface	bottom	Dissolved oxygen (mgl^{-1}) surface	bottom	Mixing depth (m)
March	26.7	24.4	8.2	**0.4**	25.0
May	26.6	25.7	7.7	5.6	40.0
July	25.4	25.1	6.1	6.0	65.0
September	26.6	24.4	9.0	**4.7**	40.0
November	27.7	25.8	11.0	**2.2**	30.0
December	25.2	24.7	7.4	**2.4**	30.0

Table 2. Temperature and dissolved oxygen from offshore deep station (70 m) in Lake Victoria. Boldface represents bottom DO levels (<5 mgl^{-1}) during stratification periods. Source, Mugidde et al. (2005).

as light in the euphotic zone is adequate during this time of the year. Recent studies have showed that the lake is continuously hypoxic below 50 m depth especially during stratified period, while areas between 25 and 50 m are more subject to frequent severe hypoxia (Figure 1b, Table 1). Even in the littoral waters, low values have been measured near the bottom (Wanink et al., 2001). For example, during a 2-year-long monthly sampling programme at an 8 m deep station, a pronounced hypoxia at the bottom (DO of 2–3 mgl^{-1}) was regularly recorded in Mwanza Gulf of Lake Victoria (Akiyama et al., 1977).

5. Introduced species and hypoxia

Prior to the current ecological changes, intense fishing and introduction of exotic species, Lake Victoria ecosystem was able to counteract hitherto perturbations (Ogutu-Ohwayo, 1990). Phytoplanktivores and detritivorous haplochromines cichlids constituted the highest number of species and biomass in the lake up to late 1980s (Figure 2a). Haplochromines were able to crop and digest algae maintaining a healthy lake ecosystem (Goldschmidt et al., 1993). Predation by Nile perch and intensive fishing led to dramatic decline of haplochromine stocks. By late 1980s, haplochromines could no longer be caught in trawls at 8-50 m depth where formerly they were numerous (Witte et al., 1992a, 1992b). Over 90% of sublittoral and deep-water haplochromine stocks declined between 1968 (Kudhogania & Cordone, 1974), and 1986 (Witte et al., 1992b). Nile perch is believed to have contributed to a bottom-up and a top-down increase in hypoxia in the lake. Nile perch, an apex piscivore, is largely responsible for the near decimation of the detritivorous and phytoplanktivorous haplochromine cichlids (Ogutu-Ohwayo, 1990, 1994). Studies have shown that during its boom between late 1960s and early 1980s, the perch fed mainly on haplochromines (Figure 3a). Between late 1980s and early 1990s, haplochromines were rare in the diet of Nile perch, whereas invertebrates (primarily *Caradina nilotica*), *R. argentea*, and juvenile Nile perch were more frequent (Figure 3b; Mkumbo & Ligtvoet, 1992). This dietary shift coincided with the dramatic decline of haplochromines (Figure 2a). Between 1997 and 2010 haplochromines became again more prominent in Nile perch diet (Figure 3c-f). The shift in diet corresponded with resurgence of haplochromines in the lake. Loss of phytoplanktivorous and detritivorous haplochromines, accelerated algal production, increased eutrophication led to increased deep water hypoxia (Goldschmidt et al., 1993). Witte et al., (1992a) argued that the impact of the Nile perch on the haplochromine cichlids was much greater than that of fishing. For example, haplochromines declined dramatically in large areas with low fishing pressure such as the Emin Pasha Gulf.

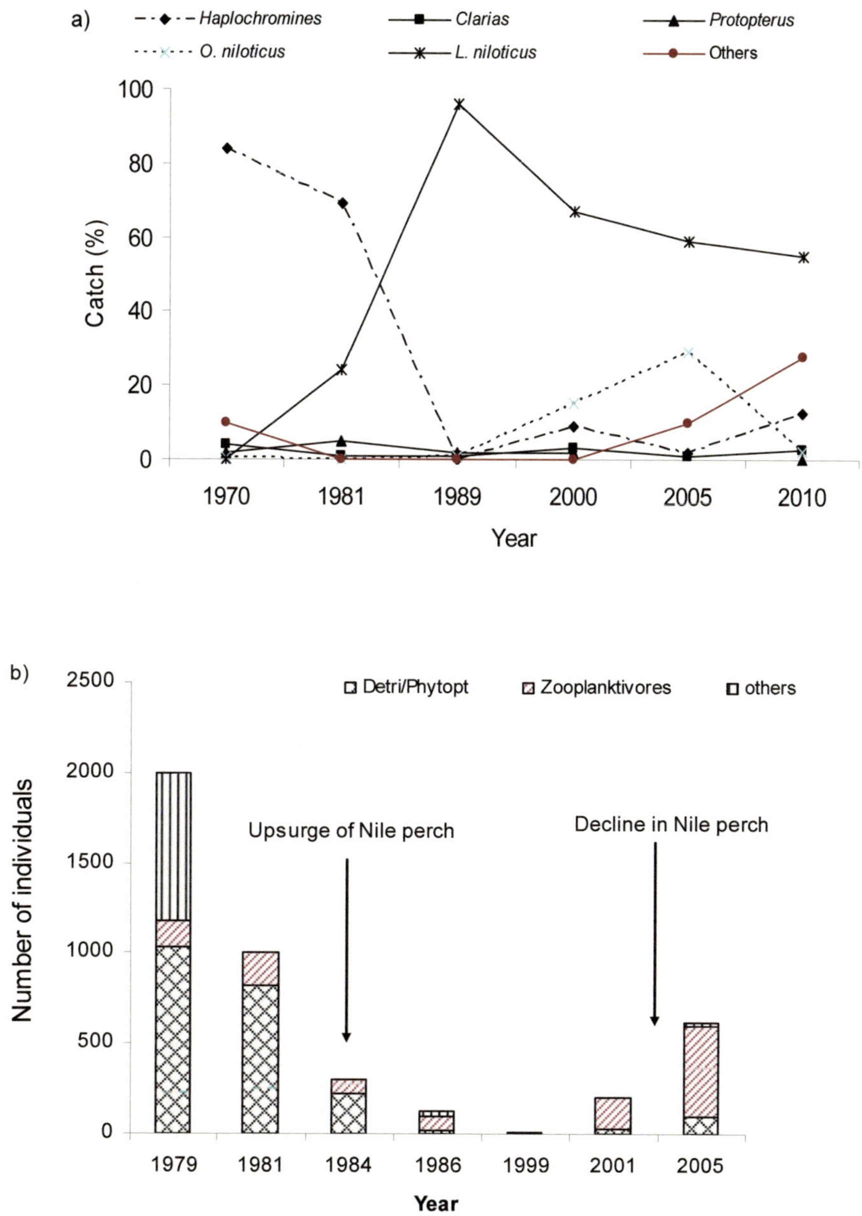

Fig. 2. Trends in haplochromine catches in Lake Victoria, a) bottom trawl catches in Nyanza Gulf, Kenya. Others = *Synodontis* spp, *Brycinus* spp. *Bagrus docmak.* Source: 1970- 2004. Njiru et al., (2005), 2010, KMFRI, b) trophic groups in Mwanza Gulf, Detri/Phytopt = detritivores/ phytoplanktivores, others = remaing trophic groupings (e.g., piscivores, insectivores, molluscivores). Source: Witte et al. (2007).

Unlike the small haplochromines, the relatively fat large Nile perch could not be dried in the sun (Riedmiller, 1994). Big fish was chopped into pieces and deep fried in their own fat while the smaller Nile perch were mainly smoked (Ligtvoet et al., 1995). For instance, at Wichlum Beach (Kenya), the number of smoking kilns increased between 1984 and 1991 from about 10 to over 50 (Riedmiller, 1994). For frying and smoking purpose, firewood used enhanced the deforestation along the lake's shore, exposing land to increased erosion and increased nutrients inputs into the lake. High nutrient levels led to high rates of photosynthesis and phytoplankton blooms created nearly continuous super-saturation in surface waters (Hecky et al., 1994). Subsequently, microbial breakdown led to reduction in DO in the bottom waters creating anoxic areas.

The exotic Nile tilapia is associated with negative effects on native tilapias (*O. variabilis, O. esculentus*) which included hybridization, overcrowding, competition for food, and possibly introduction of parasites and diseases (Trewavas, 1983; Ogutu-Ohwayo, 1990). The Nile tilapia, which is more fecund, grows faster, feeds on a variety of food items, and generally is a better competitor for resources displaced the native tilapias (Trewavas, 1983; Njiru et al., 2004).

Ecological changes in the lake, including oxygen availability, were further affected by invasion of water hyacinth, *Eichhornia crassipes* a native to South America in 1989 (Twongo et al., 1995). The weed invasion had significant socioeconomic and environmental impacts and affected water quality (Twongo et al., 1995; Njiru et al., 2002). Shading of the water by the hyacinth curtailed photosynthesis, while microbial breakdown of decaying plant material used the available oxygen. The waters below water hyacinth recorded DO as low as 0.1 mgl^{-1} making it inhabitable to most fish (Njiru et al., 2002). Additionally, the weed affected distribution of fish by blocking migratory routes of those escaping low DO and predation (Balirwa et al., 2003). However, studies by Njiru et al., (2002) found the hyacinth to have led to recovery of the native species which where more hypoxia tolerant such as catfishes, lungfish and tilapia (Figure 4).

6. Fish kills

Dissolved oxygen is one of the main factors influencing the distribution of fish in lakes (Kramer,1987; Wanink et al., 2001). In warm waters, fish generally die when exposed to DO concentrations lower than 1.5 mgl^{-1} (Miranda et al., 2000). At this concentration, the partial pressure of DO, approximately ± 35 mm Hg, is only marginally grater than venous O_2 pressure of fish, 30 mm Hg (Perry & McDonald, 1993). Fish kills is more likely from aperiodic hypoxia, with complete avoidance of persistent hypoxia (Diaz & Breitburg, 2009). Episodes of hypoxia may lead to changes in community composition, a decline in faunal diversity and decrease in fisheries production. In Lake Victoria, thermal stratification which enhances chemical reactions favouring the accumulation of toxic organic compounds and anoxic conditions has led to fish kills. Ochumba (1990) reported massive kills of Nile perch and *O. niloticus* in Nyanza Gulf of Lake Victoria which he attributed to low pH and low dissolved oxygen level, combined with physical clogging of the gills by suspended detritus and algae. The adverse conditions were as a result of massive algal blooms and their subsequent microbial breakdown in the lake. More than 80% of the dead Nile perch were adult fish (> 60 cm TL). The disproportionate number of large fish that died might have been related to a smaller gill surface area, relative to their biomass, compared to smaller fish. Moreover, smaller fish might be better at utilizing DO at the water's surface. Similarly, Fish

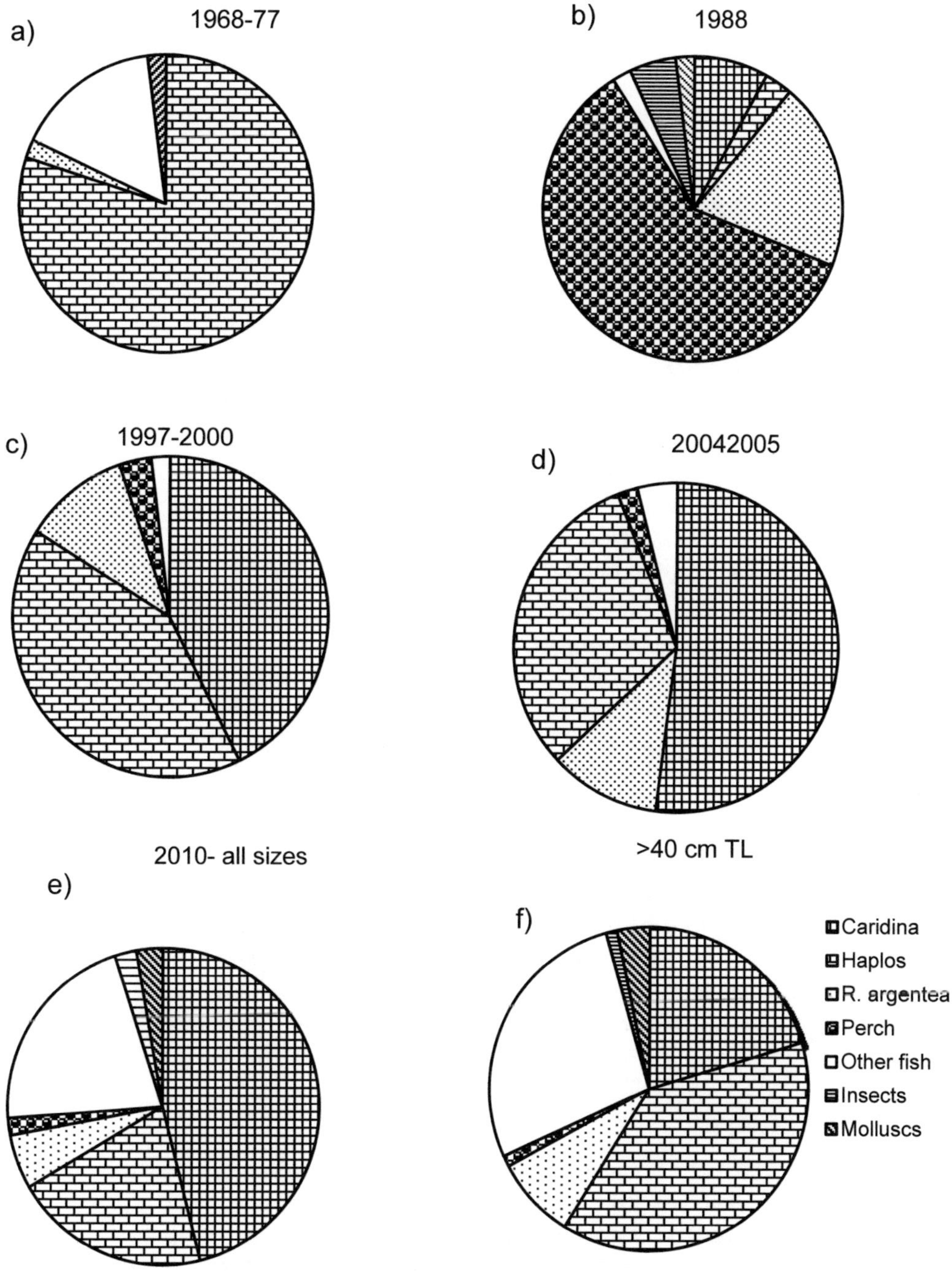

Fig. 3. Diet of *L. niloticus* in Lake Victoria; a) 1968-1977, (b) 1988, c) 1998-2000, d) 2004-2005, e) & f) 2010. Haplos = Haplochromines, Others = include *Synodontis, Clarias,* Tilapias. 1968-2005 data adapted from Njiru et al., (2005), 2010 data from KMFRI.

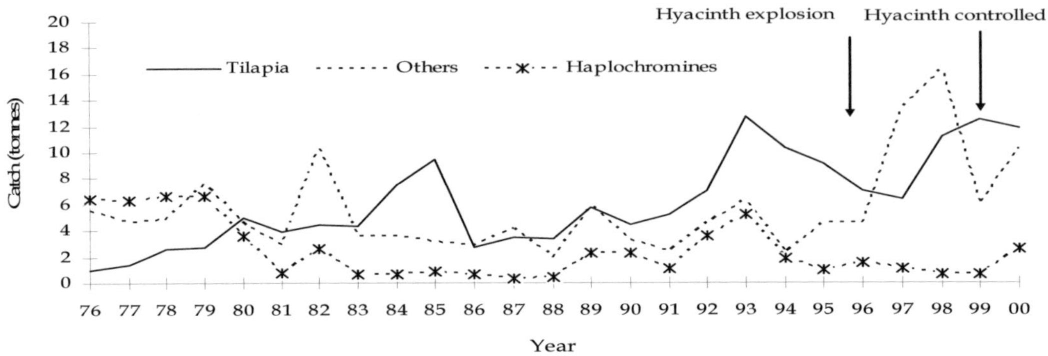

Fig. 4. Annual catches of haplochromines, tilapines and other species (mainly *Clarias Protopterus, Synodontis*) in Lake Victoria, Kenya. Source: Njiru et al., (2005).

(1956) recorded fish kills in Lake Albert which he attributed to sudden decrease in dissolved oxygen and increased carbon dioxide. Of the dead fish, 70% were adult Nile perch. He suggested that the larger fish were more dependent on optimum aeration conditions than smaller fish. Lower numbers of dead Nile tilapias were in the records probably due to its higher tolerance to low DO and habitat restriction in relatively shallow waters.

Further, Kaufman and Ochumba (1993) using a remote operated vehicle in Lake Victoria found large numbers of dead or dying *R. argentea* below the oxycline. The authors suggested that rapid shifts in the oxycline may have caused the mass mortality. In contrast to Nile perch, *R. argentea* may sink to the bottom after dying from a sudden deoxygenation of the habitat, leaving no evidence of a kill at the surface. Wanink et al., (2001) reported massive kills of fish (mainly Nile perch) in Mwanza Gulf following severe storm, a drop in the depth of the oxycline and a subsequent sudden upwelling of hypoxic water. The stomachs of most floating Nile perch were extruded probably caused by expansion of the swim bladder, which suggested that the fish had risen quickly to the surface, driven there by upwelling hypoxic water. Other causes of fish death in Lake Victoria such as Aluminium and unionized ammonium toxicity are discounted due to their low concentrations and the slightly acidic condition (Hecky et al., 1994; Wanink et al., 2001). The absence of dead lungfish, *P. aethiopicus* and *C. gariepinus* further supports the evidence that the fish were killed by hypoxia. Lungfish can breathe atmospheric air while *C. gariepinus* has an accessory organ that enables it to extract oxygen from the air in waters with low concentration.

7. Oxycline as refugia

Prey species of Nile perch may find refugia in hypoxic habitats, since this predator is thought to be extremely sensitive to low oxygen levels (Kaufman, 1992). Support for this 'hypoxic predation refugium' hypothesis came from exploratory dives by a remote operated vehicle, which filmed dagaa shoaling near the oxycline (Kaufman & Ochumba, 1993). Nile perch was in waters with a higher DO, but made short foraging grips into the hypoxic layer. According to Hecky et al., (1994) the hypoxic deep-water area may have originally functioned as a refugium for the haplochromines that had lower critical DO levels than Nile perch. The increasing deoxygenation (<1 mgl^{-1}) of the deep water might have forced the demersal populations to the shallower waters, where they were exposed to Nile perch predation. Chapman et al., (1995) further, suggested that some species that were thought to have disappeared could have sought refuge in low oxygen areas or near the oxycline in deeper waters that were not adequately sampled.

However, the ability of Nile perch to dive below the oxycline limited the use of these habitats as a refugium. Wanink et al., (2001) showed presence of Nile perch at concentrations of oxygen of less than 3 mg O_2 l^{-1} to be common (Figure 5). Wanink et al., (2001) refutes the fact that *R. argentea* benefited from low oxygen refugia. The authors found no difference in the levels at which dissolved oxygen starts to limit the distributions of *R. argentea* and Nile perch. Nile perch may even cope better with extreme hypoxia than *R. argentea*, since at levels below 1–2 mg O_2 l^{-1} the cumulative catch of Nile perch was higher than that of dagaa. In addition, it is likely that the critical oxygen level that limits Nile perch distribution in the field is lower than 5 mg O_2 l^{-1}, the level used in the hypoxic predation refugium hypothesis. This level is based on experiments by Fish (1956), who found that, in the absence of carbon dioxide, the Haemoglobin of Nile perch cannot be fully saturated unless the water in contact with the gills contains at least 5 mg O_2 l^{-1}. Up to now no studies have shown that Nile perch are limited to water in which their haemoglobin is fully saturated. Equally, bottom trawling in Mwanza Gulf of Lake Victoria yielded high Nile perch catches throughout the period 1986– 87 (Ligtvoet & Mkumbo, 1990), whereas a previous limnological study in that area had shown that the level of dissolved oxygen near the bottom was below the level of 5 mg O_2 l^{-1} for several months per year (Akiyama, et al., 1977).

Besides, haplochromine cichlids which fed on the bottom also declined dramatically in catches (Witte et al., 1992a, 1992b; Goldschmidt et al., 1993). The disappearance was attributed to hypoxia and predation by Nile perch. Verschuren et al., (2002) suggested that deoxygenation of deep water have eliminated critical haplochromine refugia, exacerbating the vulnerability of haplochromine cichlids to Nile perch predation. Haplochromine cichlids may have faced the choice of death by asphyxia in the deep anoxic water or predation by the Nile perch in the oxygen-rich shallow waters. Nonetheless, Wanink et al., (2001) noted that an increasing thickness of the hypoxic layer with distance from shore could limit movement of Nile perch. This scenario would allow hypoxia tolerant species to seek refugia and feeding grounds in the deeper waters below the oxycline. For example, invertebrate species capable of tolerating low DO may use hypoxic benthic areas in stratified lakes as refugia to avoid fish predation (Kaufman, 1992). Increasing hypolimnetic anoxia in Lake Victoria coincided with changes in the macroinvertebrate community. There has been a notable increase of the detritivorous shrimp *Caridina nilotica,* chironomids and mayflies, *Povilla adusta* (Witte et al., 1992a, 1992b; Goldschmidt et al., 1993). These invertebrates probably use low-oxygen water as refugia from Nile perch predation. They may also be exploiting increased availability of algal and detrital foods, caused by the disappearance of the haplochromines (Kaufman, 1992; Goldschmidt et al., 1993).

8. Wetlands and rocky habitats

Studies in Lake Victoria and its satellite lakes suggest that wetlands and rocky habitats play a crucial role in predator-prey interactions (Chapman et al., 1996). Prey tolerant to high hypoxia may seek refugia in wetlands which are less accessible to less tolerant predators. The well-oxygenated rocky shores and offshore rocky islands may also serve as important

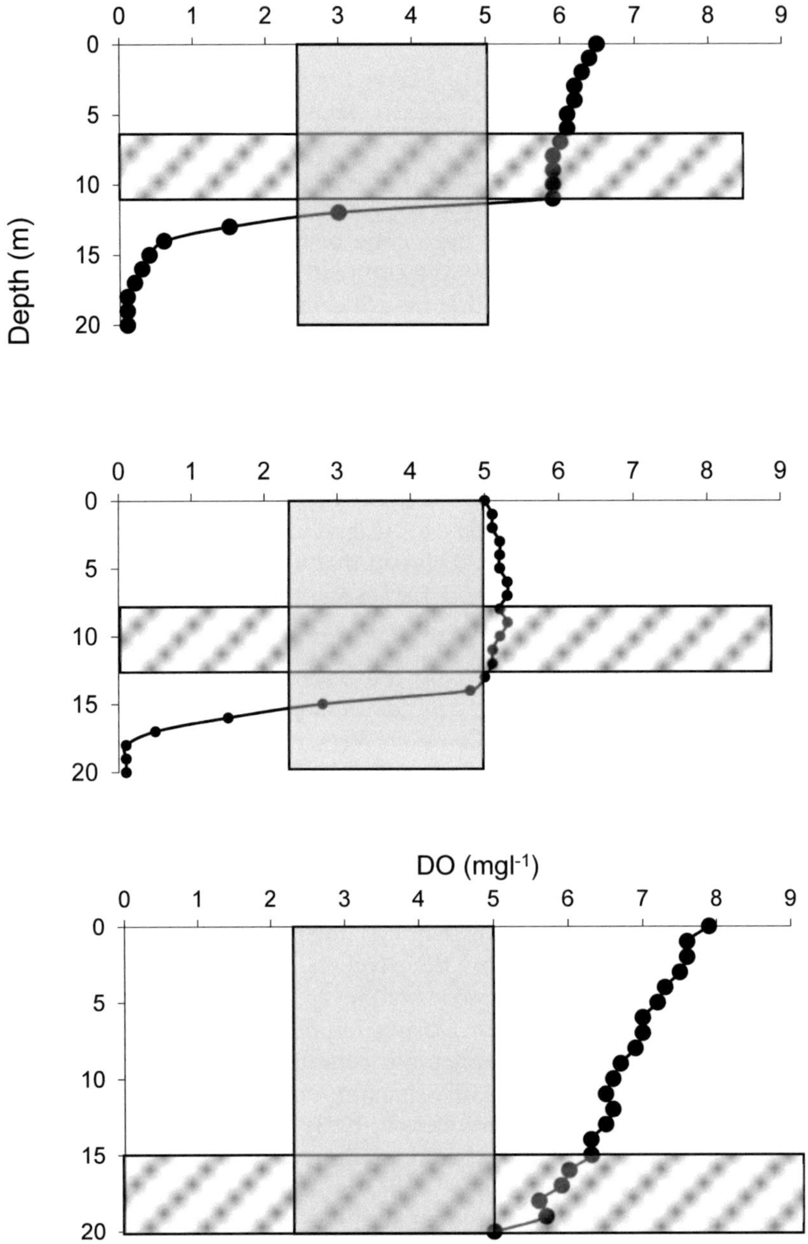

Fig. 5. Day-time depth distribution of adult *R. argentea* in relation to the depth of the oxycline in Mwanza Gulf, Lake Victoria on three different days in 1988 (from top to bottom: April 28, February 25 and October 14). Hatched bars represent levels at which *R. argentea* was caught. The vertical grey bars represent the oxygen concentration below the level needed by Nile perch to fully saturate their Haemoglobin (Fish, 1956), at which their abundance did not seem limited. Source: Wanink et al. (2001).

refugia because of their high structural complexity (Seehausen et al., 1997). Wetlands and rocky crevices are believed to have mitigated the impacts of predaceous Nile perch allowing survival of some native fish species (Witte et al., 1992a; Seehausen et al., 1997; Chapman et al., 1996). Wetlands around Lake Victoria and its satellite lakes are typically dominated by stands of *Miscanthidium violaceum* or *Cyperus papyrus* which form dense canopies. Light and wind hardly infiltrate the swamp canopy and both plant growth and decomposition proceed at high rates (Chapman et al., 1996). Gradients of very low levels of oxygen are created from the interior of the swamp to the open waters (Schofield & Chapman, 2000). Environmental conditions in wetlands are variable with DO oscillating from 0.1 mgl^{-1} to 7.0 mgl^{-1} in less than 12 hours (Chapman et al., 1998). The hypoxic conditions in wetlands function as barriers to dispersal of Nile perch (Chapman et al., 1996; 2002; Schofield & Chapman, 2000), and act refugia for more hypoxia tolerant fish species. Unlike the oxycline, where Nile perch can dive to get prey, the large areas of wetlands make it difficult for the perch to penetrate (Chapman et al., 1995; 1996; 2002; 2003; Rosenberger & Chapman, 1999).

Experimental work on Lake Victoria cichlids, initially the principal prey of the introduced Nile perch, showed generally high levels of hypoxia tolerance (Chapman et al., 1995). The authors suggested that some cichlids may have used the structural complexity of the rocky, littoral waters and fringing swamps as a defense against Nile perch. This suggestion is supported by relatively high haplochromine survival in rocky shore habitats and wetlands of Lake Victoria (Witte et al., 1992a,1992b; Seehausen et al., 1997). In Lake Nabugabo, haplochromine cichlids are more confined to inshore areas, particularly the wetland ecotones (Chapman et al., 1996; Schofield & Chapman, 2000). The overlap between Nile perch and indigenous species was greatly reduced in the ecotone areas. Juvenile Nile perch (8.6–42.2 cm, TL) were 3.7 times more abundant in offshore exposed areas than in inshore areas near wetland ecotones. Stomach analysis of Nile perch found fewer haplochromines and indigenous species in fish caught in ecotonal zones compared to exposed habitats. This suggested that the ecotone of the wetland/open water acted as a refugium probably due to higher DO and structural complexity.

However, recent studies have found a larger proportion of Nile perch near wetland ecotones with DO as low as 0.96 mgl^{-1} than in the mid 1990s (Paterson & Chapman, 2010). These juveniles penetrating the hypoxia zones had bigger gill sizes than their counterparts in the open waters. The advantage of Nile perch occupying wetland ectones was probably to maximize prey capture, and seek refuge from cannibalism and intense fishing. Although intense fishing pressure on Nile perch coincided with faunal resurgence in Lake Victoria, it is possible that other environmental changes have contributed to the observed patterns (Njiru et al., 2002). The resurgence in native species in the early 2000s coincided with the lake invasion by water hyacinth. The hypoxic water beneath the mats may have acted as a refuge for native species such as haplochromines, catfishes and lungfish that are less sensitive to low oxygen levels than Nile perch (Njiru et al., 2002). The hyacinth mats reduced fishing pressure allowing a recovery of these species. Similar observations were recorded in Lake Kyoga, a satellite of Lake Victoria following spread of water hyacinth (Ogutu-Ohwayo, 1994). Several other satellite lakes that have not experienced introduction of Nile perch still retain the original fish fauna similar to that of Lake Victoria in the 1950s (Maithya, 1998). For example, the extirpated species of *O. variabilis* and *O. esculentus* in Lake Victoria are abundant in Lake Kanyaboli in Kenya. It is argued that the wetlands, satellite lakes and rocky refugia may modulate extinction in the Lake Victoria by acting as source of resurging species (Kaufman et al., 1997; Chapman et al., 2003)

9. Behaviour adaptation to hypoxia

Fishes have evolved a variety of solutions to hypoxic stress which vary among species, life stages, and habitats (Kramer, 1987; Chapman & McKenzie, 2009). Well-documented responses to short-term hypoxia include behavioral and physiological mechanisms as avoidance, increased aquatic surface respiration, increased gill ventilation activity, reduced motor activity, and reduced metabolic scope, increased hemoglobin/hematocrit concentrations, and anaerobic metabolism (Rutjes, 2006; Pollock et al., 2007; Chapman & McKenzie, 2009).

9.1 Distribution

For fishes, the level of dissolved oxygen is one abiotic factor that can limit habitat utilization (Kramer, 1983, 1987; Chapman & Liem, 1995). Low-oxygen conditions may render the habitat inhabitable, change food availability, and act as barriers or biological filters to fish dispersal (Chapman et al., 1999). Habitat compression or habitat "squeeze" can occur where hypoxia overlaps with nursery habitats or makes deeper, cooler water unavailable (Domenici et al., 2007). Fish will be forced to seek alternative habitats or perish. Hypoxic conditions also shift the balance of the interaction to favour predator or prey depending on their relative tolerance of each (Domenici et al., 2007).

In Lake Victoria, severe hypoxic conditions (<1.0 mg^{-1}) now persists at depths below 40-50 m which cover about 35% of the lake's total bottom area (Wanink et al., 2001). Even in shallow areas (<20 m) more hypoxic habitats are now frequently detected (Figure 1b). The shallow areas of the lake are the fish breeding and nursery grounds. Limnological studies indicated that between 40- 50% of the fish habitat areas in the lake have been lost because of deoxygenation of its waters (Mugidde et al., 2005).

Studies show a relationship between catches and utilization of habitats in Lake Victoria. Recent acoustic studies showed that Nile perch catches increased with increase in oxygen concentrations (Figure 6). No fish were caught when DO was below 2.5 mgl^{-1}. These results concur with those of Wanink et al., (2001) on the same species. The authors caught no fish when DO was less than 0.2 mgl^{-1}, and catch rate increased rapidly with increased DO concentration. Bottom-trawl and gill-net surveys by Goudswaard et al., (2004), showed that zooplanktivorous haplochromines, *R. argentea* and piscivorous *S. intermedius*, dwelt near the bottom by day and migrated towards the surface at night. The trend seemed to be governed by prey availability. After the Nile perch boom, and following the decline of the haplochromines, *R. argentea* would invade the lower part of the water column by day (Wanink et al., 2001). Movement to the bottom was to exploit the increased stocks of benthic invertebrates, such as the shrimp *C. nilotica* and chironomid larvae. However, this changed following decreased levels of DO in the lake (Figure 5). When the lake is stratified, *R. argentea* spends the day just above the oxycline instead of near the bottom.

Before spread of hypoxia in the lake, the Nile tilapia stayed near the bottom during day and night (Goudswaard et al., 2004). This successful establishment of Nile tilapia in the lake was partly attributed to its ability to withstand a broad range of environmental variation (Njiru

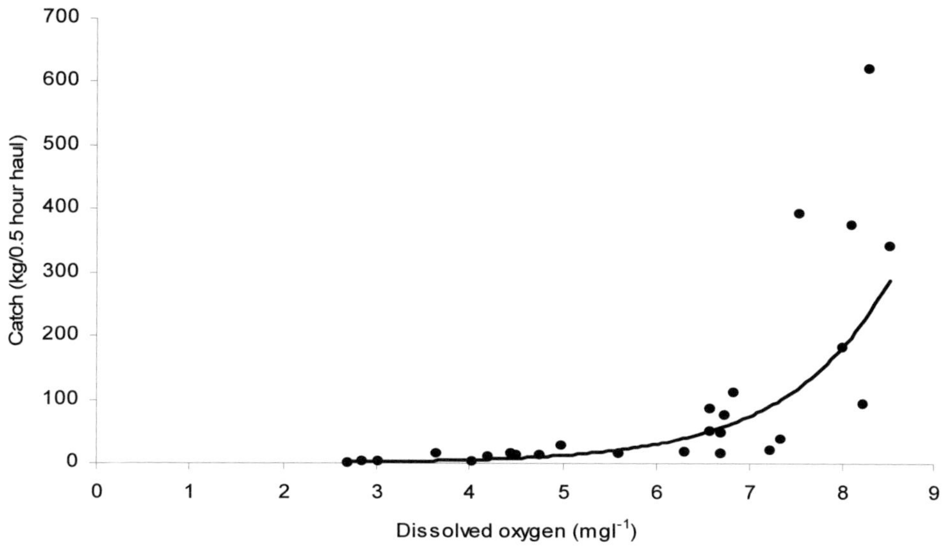

Fig. 6. Trends in Nile perch catches with oxygen in Lake Victoria, Kenya in 2010. Data from bottom trawl surveys, KMFRI 2010.

et al., 2004). Measurements of oxygen affinities in the lake for Nile tilapia (P_{50} = 1 mmHg) found the species to be hypoxic tolerant (Verheyen et al., 1986), enabling it to exploit the deeper waters and over mud bottoms. Nile tilapia juveniles are very tolerant of hypoxic stress and could even penetrate deep swamp refugia within the Lake Victoria basin (Chapman et al., 1996, 2002).

A study by Fish (1956) showed that mormyrid *Mormyrus kannume* Forsskål and the catfish *B. docmak* had P_{50} values ca. 1 mm Hg. Both species were caught up to depths of more than 60 m in the lake (Kudhongania & Cordone 1974), which are normally hypoxic. Nile perch is a bottom dweller, but it may migrate into the water column when oxygen concentrations are insufficient near the bottom (Ligtvoet & Mkumbo, 1990). The species has has been caught in Lake Victoria over a wide depth range of 1 to 60 m (Ligtvoet and Mkumbo, 1990). During its boom, the highest catch rates in the Lake Victoria were between the depths of 20 and 50 m deep (Ligtvoet and Mkumbo, 1990). A study by Fish (1956) revealed a P_{50} value of 17 mmHg for Nile perch (referred to as *Lates albertianus*), indicating that it is not hypoxia tolerant. Decreased oxygen levels in the lake have affected its distribution by confining it in areas with high DO. Recent studies revealed more hypoxia tolerant Nile perch (Paterson & Chapman, 2010). In Lake Nabugabo, the perch with large sized gills which has better oxygen uptake, is penetrating the wetland which hitherto was out of bounds for the species.

Anoxia may have forced the demersal haplochromine populations (Witte et al., 1992a,1992b; Kudhongania & Cordone, 1974) to shallower water where they are exposed to Nile perch predation. Nile perch require high concentrations of oxygen (Fish 1956), while some haplochromines can tolerate oxygen as low as 2-3 mgl^{-1} for extended periods (Van Oijen et al., 1981). The survival and eurytopic distribution of *Pseudocrenilabrus multicolor victoriae* Seegers in the lake basin is attributed to high flexibility tolerance to hypoxia (Chapman et al., 1996, 2002; Rosenberger & Chapman, 2000). The species was able to exploit hypoxic refugia such as dense wetlands which gave it an advantage over other haplochromines during the Nile perch boom period. Complete mixing of the lake occurs around June to July resulting in almost uniform distribution of nutrients and dissolved oxygen (Hecky et al., 1994; Mugidde et al., 2005). More oxygenated water allows fish to occupy deeper zones of the lake. The lake tends to have low densities of fish during mixing as fish are more widely distributed throughout the lake. The more tolerant species such the catfishes (*Clarias* , *Synodontis,* Mormyridae) and air breathing *P. aethiopicus* are found in swamps and deep waters dominated by low DO (Chapman & Liem, 1995; Chapman et al., 1996), where chronically low oxygen conditions may not limit their dispersal.

9.2 Aquatic surface respiration

Aquatic surface respiration (ASR) involves rising to the surface and ventilating the layer of water in contact with the atmosphere, which is richer in DO than the underlying bulk of water (Chapman & McKenzie, 2009). ASR is a pronounced behavioral response by bony fishes to aquatic hypoxia. This behavioral responses to hypoxia can influence other critical components of fishes in their environment, including habitat use and selection, predator–prey interactions, competitive interactions, and patterns of aggregation. A major ecological benefit to the performance of ASR is that it allows fish to colonize hypoxic refugia that less tolerant predatory species cannot occupy successfully (Chapman et al., 1995, 2002; Rosenberger & Chapman, 1999). For example, Chapman and Chapman (1998) reported remnant populations of a small mormyrid, *Petrocephalus catostoma* (Günther), in wetland lagoons surrounding Lake Nabugabo a satellite lake of Lake Victoria, after its population had disappeared from the main lake following the introduction of Nile perch. This tiny electric fish survives by virtue of a low metabolism, a low critical oxygen tension, large gill surface area, and inverted swimming during ASR to expose its subterminal mouth to the surface of the water.

Experiments on response to short-term low DO regime in cichlids from Lake Victoria (haplochromines, Nile tilapia and *O. eculentus*), revealed a well-developed ASR (Chapman et al., 1995). *Astatoreochromis alluaudi* Pellegrin which is widely spread in Lake Victoria basin showed a well-developed ASR and no loss of equilibrium during acute hypoxia (30 minutes at < 0.2 mg l^{-1} DO). In contrast, the rock-dwelling H. (*Neochromis*) *rufocaudalis* (= *N. nigricans* in Chapman et al., 1995), which normally lives in shallow well-oxygenated water, lost equilibrium more quickly than other species. When there was no access to the surface only the native tilapiine *O. esculentus* and the introduced Nile tilapia did not lose equilibrium in acute hypoxia, revealing an extreme hypoxia tolerance. Experiments by Schofield and Chapman (2000) showed that Nile perch which is intolerant of low dissolved oxygen conditions initiated ASR when oxygen tension fell below 40 mm Hg (approximately 3 mgl^{-1}). Nile perch gill ventilation rates increased with declining oxygen levels even after the initiation of ASR. For Nile perch, ASR may increase oxygen availability but not to a degree necessary to permit a lower gill ventilation rate like in other species. Work with cichlids showed a decrease in gill ventilation rates with the onset of ASR (Chapman et al., 1995). It is likely that the improvement of oxygen uptake associated with respiration at the air-water interface permitted a lower gill ventilation rate in these species.

One major physiological cost to ASR is the increased locomotor activity required for repeated surfacing and skimming (Kramer, 1987). ASR places fish at significantly greater risk from aerial predation by birds (Kramer et al., 1983). It has been demonstrated that if fish perceived a risk of predation they modulated the behavioral component of the ASR chemoreflex (Shingles et al., 2005). In Lake Victoria, *Barbus neumayeri* Fischer decreases the cost of ASR by having a more efficient oxygen-uptake method that reduces time at the surface and lower the threshold for ASR (Olowo & Chapman, 1996). Shingles et al. (2005) found that exposure of flathead grey mullet to a model avian predator delayed the onset of ASR in hypoxia or in response to direct chemoreceptor stimulation with Sodium cyanide (NaCN). Additionally, the fish surfaced preferentially under a sheltered area in their experimental chamber or close to the walls. Thus, the behavioral component of the ASR reflex is plastic; it can be modulated by inputs from higher centers, in particular as a function of perceived risk of predation. A number of studies have demonstrated decreased spontaneous activities such as; feeding rate, metabolism, and/or predator activity under hypoxic stress (Breitburg et al., 1994; 2009; Ripley & Foran, 2006). In this state, the risk of predation by piscivores may be reduced for fish prey; however, other aquatic predators may still take advantage of the negative effects of hypoxia on the ability of fish to escape (Domenici et al., 2007), thus shifting the balance towards a new player in the game.

9.3 Air-breathing

Fish breathing air use many structures to take up oxygen directly from air and some are able to switch from water to air breathing according to dissolved oxygen availability (Graham, 1997; Ilves & Randall, 2007). Just as in ASR, air-breathing is considered a behavioral response to hypoxia. A number of bony fishes have evolved bimodal respiration, retaining functional gills but can also gulp air at the water surface and store this in an air-breathing organ (ABO). The prevailing opinion is that hypoxia was the essential driving force for its evolution, and that it evolved from pre-existing ASR and bubble-holding behaviors (Graham, 1997; Graham and Lee, 2004). The primary benefit of air-breathing is that it makes oxygen uptake and aerobic metabolism entirely independent of the prevailing water oxygen availability (Graham, 1997). This enables fish to colonize extremely hypoxic regions. Persistence of a few relict species of phylogenetically ancient primitive bony and lobe-finned fishes, such as the lungfishes, is due in part to their ability to breathe air (Ilves & Randall, 2007).

Air-breathing in theory, is more energetically efficient than water-breathing for fishes, because air is richer in oxygen and requires much less effort to ventilate (Kramer, 1983, 1987; Graham, 1997). For example, only one species of obligatory air-breathing fish, the lungfish, is known from the Lake Victoria basin. The rarity of air-breathing fishes in habitats with low DO may suggest there must be significant costs to air breathing (Kramer, 1983, 1987). These costs may include increased vulnerability to aerial predation and increased energetic costs of travel to the surface (Kramer, 1983; Kramer et al., 1983; Randle and Chapman, 2004). Air-breathing could also interfere with social interactions if fish were dependent upon constant visits to the surface. Kramer et al. (1983) used a trained heron to evaluate the risk of aerial predation for air-breathing fishes and non-air breathing fishes that use ASR in response to hypoxic stress. They found that fish using ASR tended to surface at lower DO thresholds than air breathers, though surfacing time was longer. They suggested that fish using ASR may incur less risk of avian predation at moderate DO levels, but air breathers seem to have an advantage under extreme hypoxia. Fish have also been shown to reduce the risk of surfacing by selecting less risky habitats or at less risky times of the diel cycle (Shingles et al., 2005).

9.4 Haemoglobin switching

Fish that are exposed to short-term hypoxia normally react with increased ventilation, reduction of external activity, and increased aquatic surface respiration (Van den Thillart GEEJM & Van Waarde, 1985). When given time to adapt to the new environment, metabolic rates usually decrease resulting in higher hypoxia tolerance and reduced standard metabolic rate (Van den Thillart GEEJM *et al.* 1994). Many fish species adapt to hypoxia by reducing their metabolic rate and increasing hemoglobin oxygen (Hb-O_2) affinity. Pilot studies with young broods of cichlids showed that the young, when gradually acclimatized, could survive severe hypoxia in contrast with the adults. It was therefore hypothesized that early exposure results in improved oxygen transport. Rutjes et al., (2007) using split experiments, broods of *A. alluaudi, Haplochromis ishmaeli* Boulenger, and a tilapia hybrid (*Oreochromis*) were raised either under normoxia (NR; 80–90% air saturation) or hypoxia (HR; 10%air saturation). The activity of the mitochondrial citrate synthase was not different between NR and HR tilapia, but was significantly decreased in HR *A. alluaudi* and *H. ishmaeli,* indicating lowered maximum aerobic capacities. On the other hand, hemoglobin and hematocrit levels were significantly higher in all HR fish of the three species, reflecting a physiological adaptation to safeguard oxygen transport capacity. In HR tilapia, intraerythrocytic GTP levels were decreased, suggesting an adaptive increase of blood-O_2 affinity. Similar changes were not found in HR *H. ishmaeli.* In this species, however, all HR specimens exhibited a distinctly different iso-Hb pattern compared with their NR siblings, which correlated with a higher intrinsic Hb-O_2 affinity in the former. All HR cichlids thus reveal left-shifted Hb- O_2 equilibrium curves, mediated by either decreased allosteric interaction or, in *H. ishmaeli,* by the production of new haemoglobins. They concluded that the adaptation to lifelong hypoxia is mainly due to improved oxygen transport. Some species like *B. neumayeri* survived extreme hypoxic conditions in the wetlands by combination of low metabolism, high haemoglobin, and very efficient use of aquatic surface respiration (Chapman & Liem, 1995; Olowo & Chapman, 1996; Chapman et al., 1999).

10. Effects of hypoxia on activity

10.1 Parental care

A significant metabolic cost for some fish species is the energy directed to parental care (Chapman & McKenzie, 2009). Fishes that exhibit post-fertilization care have strategies ranging from simple nest guarding, to mouth brooding, to live bearing to ensure survival of their future generation (Fryer & Iles, 1972). Several oviparous fishes in Lake Victoria protect their developing young after spawning by selecting suitable nesting sites, as well as nest building and guarding the young, fanning to aerate eggs, and mouthing to clean and remove dead and diseased eggs (Wootton, 1990). In mouth-brooding tilapias and haplochromines, the parents protect the young from predators and other environmental stressors such as low dissolved oxygen by moving to more suitable microhabitats (Trewavas 1983; Chapman et al., 2002). The costs to the parent associated with carrying the young include high energetic costs of bearing the young and increased predation risk due to reduced mobility (Timmerman & Chapman, 2003). Nest-guarding fishes such as *T. zillii* use fanning to increase ambient oxygen levels around their eggs.

In mouth-brooders, the costs of parental care may be severe under hypoxia because of the increased requirements of oxygenating eggs when the parent cannot eat, and the elevated predation risks associated with any surfacing behavior. Nonetheless, high levels of parental care may be necessary under hypoxic conditions to ensure survival of the eggs and young. Wen-Chi Corrie et al., (2007) quantified the behavioral response to progressive hypoxia of the widespread mouth-brooding African cichlid, *P. multicolor victoriae*. This species responded to progressive hypoxia by performing ASR; however, brooding females showed higher ASR thresholds than males, and initiated ASR at a much higher threshold. Non-brooding females did not differ from males for any ASR threshold. A high ASR threshold in brooding females may reflect various costs such as churning behavior, which is used to move the brood inside the mouth, potentially enhancing ventilation and cleaning of the eggs and young (Keenleyside, 1991). This may add to the energy expenditure of the female, particularly under hypoxic conditions.

10.2 Social interactions

When hypoxia elicits behaviors such as ASR or air breathing it increases the risk of predation, impacts on mobility of prey, and alters the dynamics of schooling behavior. Spacing within schools allows fish to keep track of one another without colliding. Domenici et al., (2002) suggested that the fast sound pulses emitted by some fishes, which may assist in synchronous response to predators may be less effective when school volume is increased under hypoxic stress. Hypoxia may also affect sensory channels involved in fish maneuverability, and thus impair fast anti-predator manoeuvres (Domenici et al., 2007). The use of ASR in response to extreme hypoxia has been demonstrated to affect social behaviors within groups of conspecifics. For example, in their study of ASR in swamp-dwelling and open-water populations of the haplochromine cichlid *Astatotilapia* "wrought-iron," Melnychuk and Chapman (2002) found that the pre-ASR aggression rate was higher in swamp-dwelling "wrought-iron" than in the open-water populations, but the aggression rate dropped in both open-water and swamp-dwelling fish between the pre ASR and post-ASR periods. The use of ASR may impose both time and energetic constraints that reduce aggression. This could affect the development and maintenance of dominance hierarchies in cichlids and other species with complex social systems.

10.3 Swimming activity

Changes in spontaneous swimming activity have been described in a wide variety of fish groups and species when exposed to hypoxia. Behavioral responses can comprise either a reduction in activity or an increase in activity, depending upon the species and the context (Chapman & McKenzie, 2009). It has been suggested that species that reduce their activity in hypoxia tend to be demersal or bentho/pelagic, with a relatively sedentary lifestyle during which they may often encounter hypoxia in their habitat; whereas species that increase activity tend to be active pelagic schooling fishes (Domenici et al., 2000). Swimming is typically considered to represent a major component of the energy budget of active fishes, and high-intensity aerobic swimming can utilize a very significant proportion of a fish's aerobic metabolic scope (Claireaux and Lefranc¸ois, 2007). Thus, for those species that reduce levels of spontaneous swimming activity in hypoxia, this has been interpreted as an energy-saving response.

11. Morphological changes

11.1 Gill structure

The adaptive responses in fish in environments with increased hypoxia include an enlargement of the respiratory surface area (Chapman et al., 2000; Wanink & Witte, 2000; Rutjes et al., 2007; Witte et al., 2008). A larger surface area allows for a more efficient oxygen transfer (Chapman et al., 2000). Studies of morphophysiological adaptations to hypoxic conditions in haplochromine cichlids from the Lake Victoria basin have indicated that a larger gill surface area is a common response to hypoxia both within and among species (Galis & Barel, 1980; Hoogerhoud et al., 1983; Chapman et al., 2000). Offspring of adult *H. pyrrhocepalus* raised in hypoxia (0.8 mg O_2 l^{-1}) gill surface area was 80% larger than that of their full sibs raised under normoxia (6 mg O_2 l^{-1}) (Rutjes, 2006). Chapman et al. (2000) reported 18% difference in gill surface between *P. multicolor victoriae* reared in hypoxia (1.0 mg O_2 l^{-1}) and normoxia-reared sibs, and a difference of 29% for geographically isolated natural populations from a hypoxic (0.4–3.8 mg O_2 l^{-1}) and a well-oxygenated habitat (6.1 mg O_2 l^{-1}). Similar increase in gill size was found in mormyrids *Gnathonemus victoriae* (Worthington), and *P. catostoma* from dense hypoxic swamp refugia near Lake Nabugabo compared to populations from well-oxygenated waters of nearby lakes (Chapman & Chapman, 1998).The increase in gill surface area may be an adaptive response to hypoxic stress that has facilitated use of deep swamp refugia and permitted persistence with Nile perch in the Nabugabo system.

When *R. argentea* shifted towards a more benthic habitat an increase in the number of gill filaments and a decrease in the number of gill rakers was noticed. Wanink and Witte (2000) found the number of gill filaments on the first gill arch in *R. argentea,* to have increased by 3.6% between 1983 and1988. The increased gill size was postulated to have improved the capacity of *R.argentea* to extract oxygen from the water which had less oxygen than the previous habitat. Paterson and Chapman (2010) found significantly larger gills in Nile perch caught in wetland than in open-water habitats of Lake Nabugabo in 1996 and 2007. The larger gill size in wetland specimens could be used to extract oxygen in waters with lower DO levels. Witte et al., (2008) provided more direct evidence for rapid morphological changes in the Lake Victoria region by comparing archived specimens of an endemic haplochromine (*Haplochromis* (*Yssichromis*) *pyrrhocephalus* Witte & Witte-Maas) to contemporary conspecifics. The zooplanktivorous *H. pyrrhocephalus* nearly vanished coincident with the upsurge of Nile perch in the 1980s, but recovered in the 1990s. The authors reported a total gill surface area of 64% greater in recently collected specimens (1993–2001) than in conspecifics collected prior to the Nile perch explosion (1977–1981). Further, Hoogerhoud et al., (1983) found a 1.6 times increase in gill size in *Haplochromis* (*Gaurochromis*) *iris* Hoogerhoud & Witte, compared to closely related *H.* (*G.*) *hiatus.* During the rainy season, stratification and low DO (2-3 mgl^{-1}) were occasionally observed in the deeper habitat (8-15m) of *H. iris* (Van Oijen et al., 1981). The capacity for plastic changes in all these species was to provide an adaptive response to fluctuations in environmental conditions especially pertaining to levels of DO.

11.2 Head size

Heads of fish are densely packed with muscles, bones, and other structures that are necessary for respiration, vision, feeding, and other functions. As a consequence of limited space, an increase in gill size may cause spatial conflicts with surrounding structures (Chapman et al., 2000). Consequently there is need for internal reorganization and change in head shape to create more space (Chapman et al., 2000). Chapman et al. (2000) found that hypoxia-reared *P. multicolor victoriae* had a greater head size than their normoxic sibs. They attributed enlarged head size to an increase in space necessary for the larger gill apparatus. Head length, and total head volume were notably smaller in wild caught modern *H. pyrrhocephalus* that experienced increased hypoxia. Total gill surface in the resurgent *H. pyrrhocephalus* increased by 64% (Witte et al., 2008). Plasticity experiments with *H. pyrrhocephalus* from the pre-Nile perch period resulted in an increase of 9.4% in head volume of hypoxia-reared fish (Rutjes, 2006). Bouton et al., (2002a) studied the head shape of a number of closely-related haplochromine species from four rocky islands with water ranging from murky with low DO levels (3.8 mg l^{-1}) to clear with high DO levels (5.1 mg l^{-1}) in Lake Victoria. Of eight environmental variables, oxygen was the most important factor influencing head shape. Fish from sites with high DO had small narrow opercular compartment (containing the gills) compared to those from locations with low DO.

12. Phenotypic plasticity and evolution

Phenotypic plasticity (PP) an environmentally induced change in the phenotype is frequently discussed in relation to speciation (Chapman et al., 2000; Bouton et al., 2002b). Low oxygen availability has led to evolution of varied morphological and physiological plasticity in fishes, including development of breathing organs, increased gill size, modification in blood carrying capacity, and change in metabolism. Phenotypic plasticity due to hypoxia may increase the probability of fish to colonize, survive, and reproduce in changing or new environments, probably resulting in genetic change through assimilation (West-Eberhard, 1989). In addition, strong selection pressure for hypoxia tolerance may lead to geographical variation among populations. Air breathing and the ultimate evolution of terrestrial vertebrates is thought to have been an evolutionary response to low atmospheric and dissolved oxygen concentrations during the Devonian period (Clack, 2007). Several studies have documented phenotypic plasticity in the jaw, skull, gill size and body shape of fishes (Wanink & Witte, 2000; Chapman et al., 2000; Bouton et al., 2002b). There are suggestions that environmentally induced phenotypic variation can be selectively advantageous and can contribute to the origin of new traits (Bouton et al., 2002).

Resurging populations in Lake Victoria are encountering a changed environment compared to the one they lived in earlier on. It is therefore logical to expect the new fauna to differ in composition and ecosystem function from the previous. Species that shifted to hypoxic wetland refugia in response to predation by Nile perch or those that are recovering in the currently hypoxic areas of Lake Victoria have potentially experienced strong selection pressure for low oxygen tolerance over multiple generations. For these species, changes have been found in respect to gill size, body shape, and respiratory physiology (Wanink &

Witte, 2000; Witte et al., 2008). However, it is not fully known whether these changes are the result of a heritable response to selection or phenotypic plasticity, or both. Nonetheless, over several generations of strong selection pressure in hypoxic refugia it is possible that heritable changes among these characters may occur. This may lead to geographical variation between hypoxia dwelling populations and those in high DO regions, particularly in traits related to oxygen uptake.

This is particularly possible in haplochromine cichlids of Lake Victoria which are well-known for their rapid evolution and adaptive morphological radiation. More than 500 endemic species evolved within the geologically short period of 100 000 - 400 000 years. Phenotypic plasticity was documented in the feeding apparatus, in gills and in eye structures (Chapman et al., 2000; Bouton et al., 2002b; Rutjes, 2006). For example, the recovery of *H. pyrrhocephalus* the most common haplochromine in Mwanza Gulf encountered greatly changed environmental conditions (Witte et al., 2008). In the 1970s, DO levels in its habitat was rarely below 3 $mgl^{-1,}$ while after recovery in the mid-1980s the same habitat often had DO even below 1 mgl^{-1} (Wanink et al., 2001). In response to the hypoxic conditions, there were changes in gill, head and eye sizes. The morphological changes over a time span of only two decades could be the combined result of phenotypic plasticity and genetic change and may have fostered recovery of this species (Witte et al., 2008). According to Chapman et al., (2000) the difference in *P. multicolor victoriae* gill size in hypoxia and well-oxygenated habitat may be the result of a combination of inherited changes and phenotypic plasticity. Similarly, the observed interdemic variation in the gill morphology of *B. neumayeri* may be due to underlying genetic differences (Chapman et al., 1999). However, environmentally induced phenotypic plasticity and/or the interaction of genetic and environmental influences may also contribute to the observed differences in gill size.

The change in external body shape may have consequences on fish locomotory performance and competitiveness on the new individuals. Chapman et al. (2008) found significant differences in the streamline and trophic morphology of *P. multicolor victoriae* that correspond to intraspecific differences in gill size. Such morphological changes may limit the survival success of phenotypes in new habitats. Alternatively, genetic exchange among cichlids may greatly be limited because they depend on choice of mates that look phenotipically like themselves. Similarly, the large difference in gill size among the swamp and open water dwelling *B. neumayeri* has a morphological cost to swamp-dwelling which may affect competitive abilities for those that move into an open water system (Chapman et al., 1999).

Taxonomists have been cautioned on reliance on morphological characters difference in species differentiation of the old and modern populations of haplochromines cichlids (Van Oijen, 1991; Witte et al., 2008). The fast responses may reflect environmentally driven phenotypic plasticity and/or canalized genetic change, in addition to the possibility of introgression through hybridization (Streelman et al., 2004). They recommend future studies to combine methods such molecular genetics, morphometrics and colour among others in differentiation of species.

13. Conclusion

The future status of hypoxia in Lake Victoria and its consequences for fish will depend on a combination of anthropogenic factors and climate change. Human population pressure will likely continue to be the main driving factor in the eutrophication and subsequent hypoxia in the lake. Expanding population and agriculture for food will result in increased nutrient loading and increased eutrophication into the lake. Global warming, increased water temperatures, and precipitation may make the lake more susceptible to development of hypoxia through direct effects on stratification, solubility of oxygen, increased organism metabolism, and remineralization rates. Large changes in rainfall patterns are also predicted (International Panel on Climate Change [IPCC], 2007). If these changes in rainfall lead to increased runoff to Lake Victoria, stratification and nutrient loads are likely to increase and worsen oxygen depletion. Equally, if stratification decreases due to lower runoff or is disrupted by increased storm activity or intensity, the chances for oxygen depletion should decrease.

Increasing and widespread deep-water anoxia in Lake Victoria might put at risk the entire fishery (Kaufman, 1992). Increasing hypoxia could lead to further changes in species composition, population decline, fish kills, and creation of widespread "dead zones," as has happened in the Gulf of Mexico and Lake Erie (Pollock et al., 2007). In Lake Victoria, such changes could have a catastrophic effect on millions of people who are economically dependent on the lake for their livelihood. It is therefore imperative to understand the consequences of hypoxia and its effect on the fisheries. To adequately manage the lake, an understanding of natural and anthropogenic factors leading to anoxia need to be fully understood. For immediate solutions, land practices or conditions that indirectly enhance deoxygenation in the lake need to be identified to underpin water quality management. Eutrophication and associated impacts of hypoxia in Lake Victoria, may be reversed if effective nutrient management is instituted. The same has happened in eutrophic lakes following a rigorous nutritive management system (Jeppesen et al., 2005). A management system which allows a heavy, but sustainable, Nile perch fishery, may allow the coexistence of many indigenous species. There are remnant populations of fishes in the lake basin in refugias such as satellite lakes that may act as potential sources for seeds for resurgence (Kaufman & Ochumba, 1993; Seehausen et al., 1997; Balirwa et al., 2003). With proper management which integrates sustainability and conservation, it might be possible to restore Lake Victoria faunal near to its original one.

14. References

Akiyama, T., Kajumulo, A.A. & Olsen, S. (1977). Seasonal variations of plankton and physicochemical condition in Mwanza Gulf, Lake Victoria. *Bulletin of Freshwater Fisheries Research Laboratory*, 27, 49–61.

Balirwa, J., C. A. Chapman, L.J. Chapman, K. Geheb, R. Lowe-McConnell, O. Seehausen, J. Wanink, R. Welcomme, and F. Witte (2003). The role of conservation in biodiversity and fisheries sustainability in the Lake Victoria Basin. *In* Lake Victoria 2000: A New Beginning. *Bioscience* 53 (8), 703-715.

Bouton N, de Visser J, Barel C.D.N. (2002a). Correlating head shape with ecological variables in rock-dwelling haplochromines (Teleostei: Cichlidae) from Lake Victoria. *Biological Journal of the Linnean Society* 76, 39–48

Bouton N, Witte, F., & Van Alphen J.J.M. (2002b). Experimental evidence for adaptive phenotypic plasticity in a rock-dwelling cichlid fish from Lake Victoria. *Biological Journal of the Linnean Society* 77, 185–192

Breitburg, D. L., Steinberg, N., DuBeau, S., Cooksey, C., & Houde, E.D. (1994). Effects of low dissolved oxygen on predation on estuarine fish larvae. *Mar Ecol Prog Ser* 104, 235–246.

Breitburg, D. L., Hondorp, D. W., Davias, L. W., & Diaz, R. J. (2009). Hypoxia, nitrogen and fisheries: Integrating effects across local and global landscapes. *Ann Revs Mar Sci* 1, 329–349.

Calamari D., Akech M.O. & Ochumba P.B.O. (1995) Pollution of Winam Gulf, Lake Victoria, Kenya: A study case for preliminary ris assessment. *Lakes and Reservoirs: Research and Management* 1, 89-106

Chapman, L.J. & Liem, K.F. (1995). Papyrus swamps and the respiratory ecology of *Barbus neumayeri*. *Environmental Biology of Fishes* 44,183-197.

Chapman, L.J., Kaufman, L.S., Chapman, C.A., & McKenzie, F.E. (1995). Hypoxia tolerance in twelve species of East African cichlids: Potential for low oxygen refugia in Lake Victoria. *Conservation Biology*, 9, 1274-1288.

Chapman, L.J., Chapman, C.A., Ogutu-Ohwayo, R., Chandler, M., Kaufman, L., & Keiter, A. (1996). Refugia for endangered fishes from an introduced predator in Lake Nabugabo, Uganda. *Conservation Biology* 10, 554-561.

Chapman, L.J. & C.A. Chapman. (1998). Hypoxia tolerance of the mormyrid *Petrocephalus catostoma*: Implications for persistence in swamp refugia. *Copeia* 1998, 762-768.

Chapman, L.J., Chapman, C.A., & Crisman, T.L. (1998). Limnological observations of a papyrus swamp in Uganda: Implications for fish faunal structure and diversity. *Verhandlungen Internationale Vereinigung Limnologie* 26, 1821-1826.

Chapman, L.J., Chapman, C.A., Brazeau, D., McGlaughlin, B., & Jordan, M. (1999). Papyrus swamps and faunal diversification: Geographical variation among populations of the African cyprinid Barbus neumayeri. *Journal of Fish Biology* 54,310-327.

Chapman, L.J., Galis, F. & Shinn, J. (2000). Phenotypic plasticity and the possible role of genetic assimilation, Hypoxia-induced trade-offs in the morphological traits of an African cichlid. *Ecol Lett*, 3, 387-393.

Chapman, L.J., Chapman, C.A., Nordlie, F.G., &. Rosenberger, A.E. (2002). Physiological refugia: Swamps, hypoxia tolerance, and maintenance of fish biodiversity in the Lake Victoria Region. *Comparative Biochemistry and Physiology* 133 (A),421-437.

Chapman, L.J., Chapman, C.A, Schofield, P.M., Olowo, J.P., Ole Seehausen L. & Ogutu-Ohwayo, R. (2003). Fish faunal resurgence in Lake Nabugabo, East Africa *Conserv Biol*, 17(2), 500-511.

Chapman, L.J., Albert, J., & Galis, F. (2008) Developmental plasticity, genetic differentiation, and hypoxia-induced trade-offs in an African cichlid fish. *Open Evolution Journal* 2, 75-88.

Chapman, L.J. & McKenzie, D. (2009). Behavioral responses and ecological consequences. In: *Fish Physiology. Hypoxia*. Vol. 27, J.G. Richards, A.P. Farrell, & C.J. Brauner eds, pp. 26–77. Elsevier, Academic Press, San Diego, CA.

Clack, J.A. (2007). Devonian climate change, breathing, and the origin of the tetrapod stem group. *Integrated Comparative Biology* 47, 510–523.

Claireaux, G., & Lefranc¸ois, C. (2007). Linking environmental variability and fish performance: Integration through the concept of scope for activity. *Phil Trans R Soc* B 362, 2031–2042.

Crul, R.C.M. (1995). Limnology and hydrology of Lake Victoria. *Verhandlungen Internationale Vereinigung Limnologie,* 25, 39–48.

Diaz, R.J. & Breitburg, D.L. (2009). The hypoxic environment. In: *Fish Physiology. Hypoxia.* Vol. 27. J.G. Richards, A.P. Farrell, & C.J. Brauner eds, pp. 1-23. Elsevier, Academic Press, San Diego, CA.

Domenici, P., SteVensen, J. F., & Batty, R.S. (2000). The effect of progressive hypoxia on swimming activity and schooling in Atlantic herring. *Journal of Fish Biology* 57, 1526–1538.

Domenici, P., Ferrari, R. S., Steffensen, J. F., & Batty, R. S. (2002). The effects of progressive hypoxia on school structure and dynamics in Atlantic herring Clupea harengus. *Proceedings of Royal Society B,* 269, 2103–2111.

Domenici, P., Lefranc¸ois, C., and Shingles, A. (2007). Hypoxia and the anti-predator behaviour of fishes. *Phil. Trans. R. Soc. B.* 362, 2105–2121.

Farrell, A.P. & Richards J.G. (2009). Defining hypoxia: an integrative synthesis of the responses of fish to hypoxia. In: *Fish Physiology. Hypoxia.* Vol. 27. J.G. Richards, A.P. Farrell, & C.J. Brauner eds, pp. 482–503. Elsevier, Academic Press, San Diego, CA.

Fish, G.R. (1956). Some aspects of the respiration of sex species of fish from Uganda. *Journal of Experimental Biology.* 33, 186-195.

Fryer, G., & Iles, T.D. (1972). "The Cichlid Fishes of the Great Lakes of Africa: Their Biology and Evolution." Oliver and Boyd, London

Galis, F. & Barel, C.D.N. (1980). Comparative functional morphology of the gills of African lacustrine Cichlidae (Pisces, Teleostei), An ecomorphological approach. *Netherlands Journal of Zoology, 30,* 392-430

Graham, J.B. (1990). Ecological, evolutionary, and physical factors influencing aquatic animal respiration. *Am Zoo.* 30, 137–146.

Graham, J.B. (1997). *"Air Breathing Fishes: Evolution, Diversity, and Adaptation"* Academic Press, San Diego.

Graham, J.B., & Lee, H.J. (2004). Breathing air in air: In what ways might extant amphibious fish biology relate to prevailing concepts about early tetrapods, the evolution of vertebrate air breathing, and the vertebrate land transition? *Physiol. Biochem. Zool.* 77, 720-731.

Goudswaard, K.P.C , Wanink, J.H., Witte, F., Katunzi E.F.B., Berger, M.R. & Postma D.J. (2004). Diel vertical migration of major fish-species in Lake Victoria, East Africa. *Hydrobiologia* 513, 141–152.

Goldschmidt, T., Witte, F. & Wanink, J. (1993). Cascading effects of the introduced Nile perch on the detritivorous: phytoplanktivorous species in the sublittoral areas of Lake Victoria. *Conservation Biology,* 7, 686–700.

Hecky, R.E., Bugenyi, F.W.B., Ochumba, P., Talling, J.F., Mugidde R., Gophen M. & Kaufman L. (1994). Deoxygenation of the deep water of Lake Victoria, East Africa. *Limnologya and Oceanograpy,* 39 (6), (1476-1481).

Hoogerhoud, R. J. C., Witte, F. and Barel, C. D.N. (1983). The ecological differentiation of two closely resembling *Haplochromis* species from Lake Victoria (*H. iris* and *H. hiatus;* Pisces, Cichlidae). *Netherlands Journal of Zoology* 33, 283-305.

Ilves, K.L., & Randall, D.J. (2007). Why have primitive fishes survived? In: *Primitive Fishes. Fish Physiology* Vol. 26, D.J. McKenzie, C.J. Brauner & A. P. Farrell, Eds., *Fish Physiology*, pp. 516–536. Academic Press, Elsevier, San Diego, CA.

IPCC, International Panel on Climate Change (2007). *"Climate Change 2007: The Physical Science Basis."* Cambridge University Press, New York.

Jeppesen, E., Søndergaard, M., Jensen, J. P., Havens, K. E., Anneville, O., Carvalho, L., Coveney, M. F., Deneke, R., Dokulil, M. T., Foy, B., Gerdeaux, D., Hampton, S. E., et al. (2005). Lake responses to reduced nutrient loading: an analysis of contemporary long-term data from 35 case studies. *Freshwater Biology* 50, 1747–1771.

Kaufman, L.S. (1992). Catastrophic change in species-rich freshwater ecosystems: The lessons of Lake Victoria. *Bioscience*, 42, 846–858.

Kaufman, L.S. & Ochumba, P.B.O. (1993). Evolutionary and conservation biology of cichlid fishes as revealed by faunal remnants in northern Lake Victoria. *Conserv Biol*, 7, 719–730.

Kaufman, L.S., Chapman, L.J. & Chapman, C.A. (1997). Evolution in fast forward: Haplochromine fishes of the Lake Victoria region. *Endeavour*, 21, (23–30).

Keenleyside, M.H.A. (1991). Parental care. In: *Cichlid Fishes: Behaviour, Ecology and Evolution.* M.H.A Keenleyside, Ed. Chapman and Hall, New York.

Kudhogania, A.W. & Cordone, A.J. (1974). Batho-spatial distribution pattern and biomass estimation of the major demersal fishes in Lake Victoria. *African Journal of Hydrobiology and Fisheries* 3, 15–31.

Kramer, D.J. (1983). The evolutionary ecology of respiratory mode in fishes: an analysis of on costs of breathing. *Environment Biology Fish*, 9, (145-158).

Kramer D.J., Manley, D. & Bourgeois, R. (1983). The effect of respiratory mode and oxygen concentration of the risk of aerial predation in fishes. *Canadian Journal Zoology* 61, 653-665.

Kramer, D.J. (1987). Dissolved oxygen and fish behaviour. *Environment Biology Fish*, 18, 81-92.

Ligtvoet, W. & Mkumbo, O. C. (1990). Synopsis of ecological and fishery research on Nile perch (*Lates niloticus*) in Lake Victoria, conducted by HEST/TAFIRI, 35–74. In: CIFA, Report of the fifth session of the sub-committee for the development and management of the fisheries in Lake Victoria, 12–14 Sept. 1989, Mwanza, Tanzania. FAO Fish. Rep. 430, FAO. Rome.

Ligtvoet W., Mous P.J., Mkumbo, O.C., Budeba, Y.L., Goudswaard, P.C., Katunzi, E.F.B., Temu, M.M., Wanink, J.H. & Witte, F. (1995). The Lake Victoria fish stocks and fisheries. In: *Fish Stocks and Fisheries of Lake Victoria: A handbook for field observations.* F. Witte & L.T. van Densen, eds, pp. 11-53, Samara Publishing Limited, Netherlands.

Lung'ayia H.B.O., M'Harzi, A., Tackx, M., Gichuki, J. & Symoens, J.J. (2000). Phytoplankton community structure and environment in the Kenyan waters of Lake Victoria. *Freshwater Biology* 43, 529–43.

Maithya, J. (1998). A survey of the Itchthyofauna of Lake Kanyaboli and other small water bodies in Kenya: Alternative refugia for endangered fish species. *NAGA. ICLARM, Quartely* 3, 54-56.

Mkumbo O.C. & Ligtvoet W. (1992) Changes in the diet of Nile perch, *Lates niloticus* (L.) in Mwanza Gulf, Lake Victoria. *Hydrobiologia* 232, 79-83.

Mugidde, R., Gichuki, J., Rutagemwa, D., Ndawula, L., & Matovu, A. (2005). Status of water quality and its implication on fishery production, In: *The state of the fisheries resources of Lake Victoria and their management* pp106-112. Poceedings of the regional stakeholders' conference. Entebe, Uganda- LVFO, Jinja, Uganda. ISBN 9970-713-10-12

McKenzie, D.J., Hale, M., & Domenici, P. (2007). Locomotion in primitive fishes. In: *Primitive Fishes. Fish Physiology*, Vol. 26, D.J. McKenzie, C.J. Brauner & A.P. Farrell Eds.,pp. 162–224. Academic Press, Elsevier, San Diego, CA.

Melnychuk, M.C. & Chapman, L.J. (2002). Hypoxia tolerance of two haplochromine cichlids: Swamp leakage and potential for interlacustrine dispersal. *Environmental Biology of Fishes* 65, 99-110.

Miranda, L.E., Driscoll, M.P. & Allen, M.S. (2000). Transient physico-chemical micro-habitats facilitate fish survival in inhospitable aquatic plants stands. *Freshwater Biology* 44, 617–28.

Njiru, M., Othina, A., Getabu, A., Tweddle, D. & Cowx I.G. (2002). "The invasion of water hyacinth, *Eichhornia crassipes* Solms (Mart.), a blessing to Lake Victoria fisheries". In: *Management and Ecology of Lake and Resevoirs Fisheries,* I. G. Cowx Ed, pp 255-263, Fishing News Books, Blackwell Science, Oxford, UK.

Njiru M., Okeyo-Owour, J.B., Muchiri M. & Cowx, I.G. (2004). Shift in feeding ecology of Nile tilapia in Lake Victoria, Kenya. African. *Africa Journal ofJ Ecology* 42, (163-170).

Njiru, M., Waithaka, E., Muchiri, M., van Knaap, M. & Cowx, I.G. (2005). Exotic introductions to the fishery of Lake Victoria: What are the management options? *Lakes and Reservoirs: Research and Management* 10, 147-155.

Ochumba P.B.O. (1990). Massiv fish kills within the Nyanza Gulf to Lake Victoria, Kenya. Hydrobiologia 208, 93-99.

Olowo, J.P. & Chapman, L.J. (1996). Papyrus swamps and variation in the respiratory behaviour of the African fish Barbus neumayeri. *African Journal of Ecology* 34, 211-222.

Ogutu-Ohwayo, R. (1990). The decline of the native fishes of lakes Victoria and Kyoga (East Africa) and the impact of introduced species, especially the Nile perch, *Lates niloticus* and the Nile tilapia, *Oreochroms niloticus*. *Environmental Biology of Fishes* 27, 81–96.

Ogutu-Ohwayo, R. (1994). Adjustments in fish stocks and in life history characteristics of the Nile perch, *Lates niloticus* L. in lakes Victoria, Kyoga and Nabugabo. Ph.D thesis. University of Manitoba. 213 p.

Paterson, J.A. & Chapman, L.J. (2010). Intraspecific variation in gill morphology of juvenile Nile perch, Lates niloticus, in Lake Nabugabo, Uganda. *Environmental Biology of Fishes* 88,97-104 DOI: 10.1007/s10641-010-9600-6.

Perry, S.F. & McDonald, G. (1993). Gas exchange. In: *The Physiology of Fishes,* D. H. Evans ed., pp. 251–78. CRC Press, Boca Raton, USA.

Pollock, M.S.,Clarke, L.M.J. & Dube M.G. (2007). The effects of hypoxia on fishes: from ecological relevance to physiological effects. *Environment Reviews* 15, 1–14.

Randle, A.R. & Chapman, L.J. (2004). Habitat use by the air-breathing fish Ctenopoma muriei: Implications for costs of breathing. *Ecology of Freshwater Fish* 13, 37-45.

Reardon, E.E. & Chapman, L.J. (2010). Energetics of hypoxia in a mouth-brooding Cichlid: Evidence for interdemic and developmental effects. *Physiological and Biochemical Zoology* 83(3), 414–423. DOI: 10.1086/651100

Riedmiller, S. (1994). Lake Victoria fisheries: the Kenyan reality and environmental implications. *Environmental Biology of Fishes.* 39, 329-338.

Rutjes, H.A. (2006). Phenotypic responses to lifelong hypoxia in cichlids Ph.D., Dissertation, Leiden University, Netherlands.

Rutjes, H.A., Nieveen, M.C., Weber, R.E., Witte, F., & van den Thillart, G.E.E.J.M.. (2007). Multiple strategies of Lake Victoria cichlids to cope with lifelong hypoxia include haemoglobin switching. *American Journal of Physiology* 293, 1376–1383.

Ripley, J.L., and Foran, C.M. (2006). Influence of estuarine hypoxia on feeding and sound production by two sympatric pipefish species (Syngnathidae). *Marine Environment Research* 63, 350–367.

Rosenberger, A.E. & Chapman, L.J. (1999). Hypoxic wetland tributaries as faunal refugia from an introduced predator. *Ecology of Freshwater Fish* 8, 22-34.

Rosenberger, A.E. & Chapman, L.J. (2000). Respiratory characters of three haplochromine cichlids: Implications for persistence in wetland refugia. *Journal of Fish Biology* 57, 483-501.

Seehausen O., Witte, F., Katunzi, E.F., Smits, J. & Bouton, N. (1997). Patterns of the remnant cichlid fauna in southern Lake Victoria. *Conservation Biology,* 11, 890–904.

Schofield, P.J. & Chapman, L.J. (2000). Hypoxia tolerance of introduced Nile perch: Implications for survival of indigenous fishes in the Lake Victoria Basin. *African Zoology* 35, 35-42.

Shingles, A., McKenzie, D.J., Claireaux, G., & Domenici, P. (2005). Reflex cardioventilatory responses to hypoxia in the flathead grey mullet (Mugil cephalus) and their behavioural modulation by perceived threat of predation and water turbidity. *Physiol. Biochem. Zool.* 78, 744–755.

Streelman, J.T., Gmyreck, S.L., Kidd, M.R., Kidd, C., Robinson, R.L., Hert, E., Ambali, A.J., & Kocher, T.D. (2004). Hybridization and contemporary evolution in an introduced cichlid fish from Lake Malawi national park. *Molecular Ecology* 13, 2471–2479.

Talling, J.F. (1966). The annual cycle of stratification and phy- toplankton growth in Lake Victoria (East Africa). *Int. Rev. Gesamten Hydrobiol.* 51, 545-621.

Timmerman, C.M. & Chapman, L.J. (2003). The effect of gestational state on oxygen consumption and response to hypoxia in the sailfin molly (Poecilia latipinna). *Environmental Biology of Fishes* 68, 293-299.

Trewavas, E. (1983). Tilapiine species of the Genera *Sarotherodon, Oreochromis* and *Danakila.* London British Museum (Natural History) Publication No. 583.

Twongo, T.F., Bugenyi, F.W. & Warda, F. (1995). The potential for further proliferation of water hyacinth in Lake Victoria, Kyoga and Kwania and some urgent aspects for research. *African Journal of Tropical Hydrobiology and Fisheries* 6, 1-10.

Van den Thillart, GEEJM., Dalla Via, J., Cattani, O., De Zwaan, A. (1994). Influence of long-term hypoxia exposure on the energy metabolism of *Solea solea*. I. Critical O_2 levels for aerobic and anaerobic metabolism. *Marine Ecology Progressive Series* 104, 109–117,

Van den Thillart, GEEJM, Van Waarde, A. (1985). Teleosts in hypoxia: aspects of anaerobic metabolism. *Mol Physiol* 8, 393–409.

Van Oijen, M.P.J., Witte F. & Witte-Maas E.L.M. (1981). An introduction to ecological and taxonomic investigations on the haplochromine cichlids from the Mwanza Gulf of Lake Victoria. *Netherlands Journal of Zoology* 31, 149–174.

Van Oijen M.P.J. (1991). A systematic revision of the piscivorous haplochromine Cichlidae (Pisces, Teleostei) of Lake Victoria (East Africa). Part 1. *Zoologische Verhandelingen* 272, 1–95.

Verheyen, E., Blust, R. & Decleir W. (1986). Hemoglobin heterogeneity and the oxygen affinity of hemolysate of some Victoria cichlids. *Comprative Biochemistry and Physiology* 84(A), 315-318

Verschuren, D., Johnson, T.C., Kling, H. J., Edgington, D.N., Leavitt P.R., Brown E.T., Talbot M.R. & Hecky R.E. (2002). History and timing of human impact on Lake Victoria, East Africa, 7 *Proceedings of Royal Society London,* B 269, 289–294. DOI 10.1098/ rspb. 2001. 1850

Stager, J.C. & Johnson, T.C. (2007). The late Pleistocene desiccation of Lake Victoria and the origin of its endemic biota. *Hydrobiologia* 596: doi:10.1007/210750-007-9158-2.

Wanink, J.H. & Witte, F. (2000). Rapid morphological changes following niche shift in the zooplanktivorous cyprinid *Rastrineobola argentea* from Lake Victoria. *Netherlands Journal of Zoology* 50 (3), 365-372.

Wanink, J.H, Kashindye, J.J, Goudswaard, P.C., Witte, F. (2001). Dwelling at the oxycline: does increased stratification provide a predation refugium for the Lake Victoria sardine *Rastrineobola argentea*? *Freshwater Biology* 46, 75–85.

Wen-Chi Corrie, L., Chapman, L.J. & Reardon, E. (2007). Brood protection at a cost: Mouthbrooding under hypoxia in an African cichlid. *Environmental Biology of Fishes* 82, 41-49.

West-Eberhard M.J. (1989). Phenotypic plasticity and theorigins of diversity. *Annual Review of Ecology and Systematics* 20, 249–278.

Wetzel, R.G. (2001). "*Limnology. Lakes and Rivers Ecosystems.*" Academic Press, San Diego.

Witte, F., Goldschmidt ,T., Goudswaard, P.C., Ligtvoet, W., Van Oijen, M.J.P. & Wanink, J.H. (1992a). Species extinction and concomitant ecological changes in Lake Victoria. *Netherlands Journal of Zoology,* 42, 214–232.

Witte, F., Goldschmidt, T., Wanink, J., Van Oijen, M., Goudswaard, K., Witte-Maas, E. & Bouton, N. (1992b). The destruction of an endemic species flock: quantitative data on the decline of the haplochromine cichlids of Lake Victoria. *Environmental Biology of Fishes,* 34, 1–28.

Witte, F., Msuku, B.S, Wanink, J.H., Seehausen, O., Katunzi, E.F.B., Goudswaard, P.C., Goldschmidt, T. (2000). Recovery of cichlid species in Lake Victoria: an examination of factors leading to differential extinction. *Rev Fish Biol Fish* 10, 233–241.

Witte, F., Wanink, J.H. & Kishe-Machumu, M. (2007). Species distinction and the biodiversity crisis in Lake Victoria. *Transaction of the American Fisheries Society*, 136, 1146-1159.

Witte, F., Welten, M., Heemskerk, M., van der Stap I, Ham, L., Rutjes, H. & Wanink, J. (2008) Major morphological changes in a Lake Victoria cichlid fish within two decades. The Linnean Society of London, *Biological Journal of the Linnean Society*, 94, (41–52).

Wootton, R. J. (1990). "*Ecology of Teleost Fishes*." Chapman and Hall, New York.

Worthington, E.B. (1930). Observations on the temperature, hydrogen-ion concentration, and other physical conditions of the Victoria and Albert Nyanzas. *Inter Rev Ges Hydrobiol*, 24, 328–357.

Part 4

Plant Methodology

6

Connection Between Structural Changes and Electrical Parameters of Pea Root Tissue Under Anoxia

Eszter Vozáry[1], Ildikó Jócsák[1], Magdolna Droppa[1] and Károly Bóka[2]
[1]Corvinus University of Budapest
[2]Eötvös Lóránd University
Hungary

1. Introduction

Anaerobiosis in plants is common. Plant species have different sensitivity to oxygen deprivation. Plants evolved anatomical or metabolic adaptations to compensate for flooded conditions are called wetland plants. Most agricultural plants tolerate waterlogging poorly and can be categorized as flood-tolerant (wheat, maize, oat, potato) or flood-sensitive (pea, tomato, soybean) plants.

The most immediate effect of anaerobic soil conditions on plants is a reduction in aerobic respiration in roots. Root cells switch to anaerobic respiration which is much less efficient than aerobic respiration. Under normal aerobic conditions 30 to 32 mol ATP are produced per hexose sugar by oxidative phosphorylation, citric acid cycle and glucolysis, whereas at anoxic conditions only 2 mol ATP per mol hexose sugar can be produced by glucolysis, which finally leads to nutrient deficiencies (Niki & Gladish, 2001). The principal end products of glycolysis are lactate and ethanol. Lactate production lowers the cytoplasmic pH and in flood-sensitive species, as pea the cytoplasmic acedosis causes cell death (Tuba & Csintalan, 1992).

In the oxygen-lack circumstance not only the metabolism of root, but that of the whole plant is sustained, because the root is not able to take up and to transfer the nutrients. There is generated ions-lack in leaves and in sprouts. In plants under long-time flood there can be appeared symptoms pointing on drought, too. Water permeability of root is decreased causing the decrease of water potential, even at closed stoma (Tuba & Csintalan, 1992).

In those plants, that genetically do not have aerenchymas, sudden flood increases the activity of cellulases and this leads to aerenchima formation in roots and vascular cavities may appear in the vascular cylinder (Niki & Gladish, 2001; Grichko & Glick, 2001).

Sarkar et al. (2008) investigated the vascular cavity formation induced by sudden flooding in pea primary roots. In the central vascular cavity of pea primary roots undifferentiated parenchymatous cells degenerate under hypoxic conditions created by flooding at temperatures higher, than 15 °C. They form a long vascular cavity that seems to provide a conduit for longitudinal oxygen transport in the roots. Specific changes in the cell wall ultrastructure were accompanied with cytoplasmic and organellar degradation in the cavity-forming roots. The degenerating cells had thinner primary cell walls, less electron-dense middle lamellae, and less abundant cell wall homogalacturonans in altered patterns compared to healthy cells of root. Cellular breakdown and changes in wall ultrastructure, however, remained confined to cells within a 50 µm radius around the root centre, even after full development of cavity. Cells farther away maintained cellular integrity and had signs of wall synthesis, perhaps from tight regulation of wall metabolism over short distances. Cold, flooded or warm, nonflooded conditions do not induce vascular cavity formation, but can also induce variations in cell wall structure (Sarkar et al., 2008).

Electrical impedance is a complex resistance in the presence of alternating current, and can be a useful tool for investigation of structural characteristics of solid materials and also animal, plant and human tissue (Macdonald, 1987; Grimnes & Martinsen, 2000). The electrical impedance spectrum of living tissue in low frequency range - from 10 Hz up to 10 MHz - can be described with various model circuits, elements of which can represent the electrical resistance and capacitance of different cellular structures (Hayden et al., 1968; Privé & Zhang, 1996; Zhang & Willison, 1993).

Different models have been created in order to approximate cellular compartments: Hayden model, in which the apoplasmic (R_a) and symplasmic (R_s) resistance is calculated (Hayden et al., 1968); the modified Hayden model also considers the cell membrane capacitance (Vozáry et al., 1999); and the "double shell model" includes the interior resistance of vacuoles and the plasma membrane capacitance as well (Zhang & Willison, 1991).

Several stress-induced alterations have already been followed in plants by electrical impedance spectroscopy. During dehydration of potato (*Solanum tuberosum* L.) and carrot (*Daucus carota*) pieces, the ratio between R_s and R_a decreases (Toyoda et al., 1994). Cold acclimation was found to enhance the impedance of alfalfa (*Medicago sativa* L.) (Hayden et al., 1968). Correlation was found between the phase angle data and the rate of aging of plasma membrane in the control and in calcium and phosphorous deficient subterranean clover (*Trifolium subterraneum* L.) (Greenham et al., 1972). Membrane injuries caused by temperature extremities can also be followed by impedance spectroscopy (Repo et al., 2000). A linear correlation was found between intracellular resistance and frost hardening capability of Scots pine (*Pinus sylvestris* L.) (Repo et al., 2000). Prediction of mechanical destruction in apple tissue under the skin is also possible with impedance spectroscopy (Vozáry et al., 1999; Vozáry & Benkő, 2010). The parameters of impedance spectrum can be used in describing the drying process of apple slices (Mészáros et al., 2005), or withering of lettuce (Vozáry et al., 2009). The germinability of soybean and snap bean seed can be predicted using the impedance parameters (Vozáry et al., 2007).

On the basis of results described in preceding paragraphs, structural changes caused by various environmental effects can be detected by electrical impedance spectroscopy, therefore the detection of structural changes as a consequence of anoxia caused by flooding of plants can also be realised. The impedance spectroscopy analysis requires neither physical nor chemical preparation of investigated plants. The measurement of an impedance spectrum takes time no more some minutes. The fastness and the simplicity of this method take competitive it compared with new microscopic techniques. This technique gives permission for detection of early changes both in physiological state and in structure following the flooding - the appearance of total lack of oxygen. The impedance spectroscopy can also be used to detect the effect of flooding in very young seedlings, when other techniques - for example fluorescence induction spectroscopy can not be realized because of the lack of fully opened leaves.

The aim of this work is to summarize the change of impedance parameters in root tissue of pea seedling of various ages under flooding stress, in anoxic condition (total lack of soil oxygen). There can be expected an alteration of the resistance in both apoplasmic and symplasmic electrolyte systems and the membrane capacitance by which stress development can be followed by electrical impedance spectroscopy that is a fast method carried out without the grinding and processing of plant tissues.

2. Materials and method

The flood-caused changes in impedance parameters of three and ten days old pea seedlings, which easily grow in artificial circumstances, were followed up to eleven days and up to ten days, respectively. The structural changes in the case of three days old seedlings were detected by light microscopy, too.

2.1 Plant material and growth conditions

Seeds of pea (*Pisum sativum* L. cv. 'Debreceni világos') were surface sterilized in 3% (w/v) sodium hypochlorite, then rinsed several times. Then seeds were soaked in distilled water for six hours after which they were sowed into a tray filled with formerly sterilized damp perlite. Germination took place in a dark room in room temperature. Two serial of experiments were executed: 1. three days old and 2. ten days old seedlings were transferred to half Hoagland solution. Half Hoagland solution consist of $Ca(NO)_3 \cdot 4\ H_2O$ (2.17 mM), KNO_3 (4.89 mM), KH_2PO_4 (1.5 mM), $MgSO_4 \cdot 7\ H_2O$ (0.134 mM), Na-FeEDTA (0.134 μM), H_3BO_4 (69.3 μM), $MnSO_4$ (13.6 μM), $ZnSO_4 \cdot 2\ H_2O$ (1.1 μM), $NaMoO_4 \cdot 2\ H_2O$ (0.15 μM),$CuSO_4 \cdot 5\ H_2O$ (0.48 μM) (Hoagland & Arnon, 1950), which was changed every third day. The seedlings were grown in a Conviron S10 type phytothrone (120 μM $m^{-2} \cdot s^{-1}$ light intensity, 22 - 23 °C, 85% relative humidity). The duration of light was 16 h.

In the case of three days old seedlings there were two serial experiments without and with aerating the hydroponic solution. The flooded ten days old seedlings were not aerated. The length of root was measured for three days old seedlings flooded, aerated and only flooded.

In each experiment 15-20 seedlings were investigated, and experiments were repeated three times. The average values with standard deviation are shown in graphs.

2.2 Morphological and light microscopic investigations

Morphological (root lengths) and tissue structure evaluation were carried out in parallel for each experiment with three days old seedlings. For light microscopic investigations, samples were collected from the epicotyl and the upper part of the primary root of the seedlings and fixed in 4% formaldehyde solution (in 0.1 M phosphate buffer, pH 7.2). After dehydration through a series of increasingly concentrated solutions of ethanol, as an intermediary solvent, benzene was used before paraffin embedding. Sections (10 μm thick) were prepared with a Leitz Wetzlar microtome and mounted on egg albumin mountant covered glass slides. Roots and epicotyls of 5 plants per treatment were sampled, from each of them 4 slides were prepared (25 sections per slides). Sections were stained with Bismarck brown and malachite green (Ruzin, 1999). Micrographs were taken with an Opton III photomicroscope.

Vascular cylinder of the pea primary root consists of tracheary elements mainly. Their number and size are influenced alike by developmental and stress-induced processes (Luxova, 1995; Iijima & Kato, 2007). The state of the vascular cylinder can be characterised by the numbers of tracheae and size of the vascular cylinder, so the product of the tracheae number (T) and the diameter of the vascular cylinder (VC) was calculated as an informative parameter of the VC wideness and cell size correlation. The average of VC diameter in a root sample was established from the maximal and the minimal diameter measured in it. Measurement of VC perimeter or estimation of square area of VC and tracheary elements are time-consuming and not more precise then product from T and VC, which is simplified but represents the main factors influencing VC structure. Vascular cavity formation was not measured separately because its presence is indicated by the product of VC and T.

2.3 Electrical impedance measurements

2.3.1 Theory of electrical impedance spectroscopy

The application of a monochromatic voltage signal at single frequency to a sample results in a current of the same frequency. If the sample has no capacitance or inductance, the phase of the current is the same as that of the voltage. However, if the sample has any capacitive reactance, a phase shift does occur. The electrical impedance (Z) of a sample is the quotient of the alternating voltage ($U = U_0 \sin \omega t + jU_0 \cos \omega t$) across the sample and the alternating current $I = I_0 \sin(\omega t + \varphi) + jI_0 \sin(\omega t + \varphi)$ through the sample. U_0 *and* I_0 are the peak values of voltage and current and U and I are the complex values of voltage and current at time t. $j = \sqrt{-1}$ is the imaginary operator and $\omega = 2\pi f$, where f is the measuring frequency. Because of this phase shift (φ) between these two sinusoidal quantities, the quotient $Z = U / I$ must be handled as a complex number, having both real (R) and imaginary (X) parts:

$$Z = \frac{U}{I} = |Z| \cdot \cos \varphi + j|Z| \cdot \sin \varphi = R + jX \qquad (1)$$

Therefore, impedance can be plotted in the complex plane, which, in this case, is also called a Wessel diagram (Grimnes & Martinsen, 2000; Macdonald, 1987). $|Z|$ and φ are the magnitude and the phase angle of impedance, respectively.

The electrical impedance spectrum is the distribution of impedance values in the frequency range of an applied alternating current. The intercellular (apoplastic) part and the intracellular (symplastic) part of living tissue has mainly resistive characteristics, while the cell membrane have capacitance with very low conductance. In living tissues, an alternating electric field causes polarization and relaxation of dipoles. At low frequency, the current passes through the apoplastic space of tissue where ions are the main current carriers. Cell membranes and other interfacial layers become conductive with increasing frequency, with the apparent current being dependent on the permittivity of interfacial layers (Pething & Kell, 1987). Accordingly, symplastic space becomes more conductive at higher frequencies. At high frequencies the symplastic and the apoplastic resistance form a parallel circuit (Zhang & Willison, 1991; Grimnes & Martinsen, 2000).

In some instances of other studies, the impedance spectrum measured on biological tissues has been modeled with equivalent circuits. In the simplest case, the measured impedance of tissue consisting of uniform, homogeneous cells can be described with a lumped model containing discrete resistors and capacitors (Harker & Maindonald, 1994; Zhang & Willison, 1991; Vozáry et al., 1999). In general, these resistors and capacitors can represent the resistance and capacitance of cell compartments, and the electrical behavior of tissue can be explained using the properties of these models (Grimnes & Martinsen, 2000). When biological tissues are non-homogeneous, distributed models are more appropriate (Repo & Zhang, 1993; Vozáry et al., 2007). If the sample under investigation contains several components having a variety of differing structures, then the measured impedance is the resultant of each structure. The equivalent lumped model of such a structure contains too many parameters for a practical mathematical description. However, these resultant impedance characteristics can be modeled using distributed elements containing fewer parameters. The trade-off, however, is that these parameters taken individually do not necessarily represent actual anatomical structures of sample tissue.

2.3.2 Measurement of electrical impedance spectra

The impedance spectra were measured with a two-pin electrode arrangement. The thickness and the length of the gold-plated copper pins were 0.35 mm and 6 mm, respectively. The electrodes were connected to BNC jacks by copper wires with Teflon isolation and the BNC jacks were joined to the LCR meter by a special 1 m length shielded cable (HP-16048A). The electrode spacing was 2 mm.

The impedance spectra were measured in roots tissues. The root was punctured with electrodes about 2 cm and 3 cm below the basis along the longitudinal axis for three and ten days old seedlings, respectively. Impedance spectrum of three days old seedlings was measured before flooding and 2, 4, 8 and 11 days after flooding. For the ten days old seedlings the impedance spectrum was determined four days daily before flooding and 1, 3, 6, 12, 24, 48, 96 and 240 hours after flooding.

The magnitude and phase angle of electrical impedance between the two pins were measured by a HP 4284A Precision LCR meter at 100 frequencies in the frequency range from 30 Hz up to 1 MHz. The input voltage level of the sine signal was 1 V. The LCR meter was connected to a computer via a GPIB interface. Data collection was performed by a QBasic program. Open (measuring air) and short (measuring short circuit) correction were applied to minimize measurement errors: stray capacitance and inductance as it is recommended by the manufacturer of the LCR meter (Honda, 1989).

2.3.3 Modelling of electrical impedance spectra

The corrected impedance spectra were approached with the modified Hayden model (Fig. 1) using the complex least square method. The modified Hayden model is a relatively simple - so called explanatory (Grimdes & Martinsen, 2000) - model of plant tissue. It contains the resistance of apoplasm (intercellular part), the resistance of symplasm (intracellular part) and the complex capacitance of cell membrane. This model describes the two parallel ways of current in living tissue: one of them is between the cells and other through the cell membrane and inside of cells.

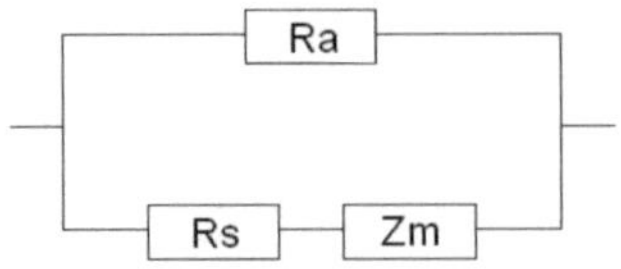

Fig. 1. Modified Hayden model. R_a, R_s are resistance of apoplasm (intercellular space) and symplasm (intracellular space) and Z_m is the impedance of membrane.

The complex electrical impedance of root tissue (Z_r) can be expressed as the resultant of circuit elements:

$$Z_r = \frac{R_a(R_s + Z_m)}{R_a + R_s + Z_m}. \quad (2)$$

Zm can be expressed as follows:

$$Z_m = \frac{\cos\psi + j\sin\psi}{C_m\omega}, \quad (3)$$

where $j = \sqrt{-1}$, $\omega = 2\pi f$, C_m is the cell membrane capacitance, and ψ is a constant phase angle, f is the measuring frequency. At low frequencies - when Z_m is very high - the impedance of tissue is equal practically with the resistance of intercellular space (R_a) and at high frequencies - at low Z_m - Z_r is the parallel resultant of R_a and R_s.

From the measured impedance after open-short correction the magnitude, $|Z_r|$, and the phase angle, ϕ_r, of pea root impedance were determined. Approaching the root impedance (Z_r) by complex non-linear squares with MATLAB program the resistance of extracellular space (apoplasm), R_a, the resistance of intracellular space (symplasm), R_s, and the cell membrane capacitance, C_m, were evaluated.

2.4 Statistical analysis

The values are expressed as the mean ±S.D. of three independent replicates of each experiment. In each experiment the impedance was measured on 15-20 seedlings. The significance for the differences of the impedance parameters between control and flood treated plants was determined by ANOVA (in Duncan multiple comparison p>0.05) using SPSS 7.0 program.

3. Results and discussion

3.1 Morphological changes of pea roots

Root length of flooded seedlings remained shorter according to the length of root of flooded and aerated seedlings (Fig. 2). This observation can be explained by the reduced transfer of nutriments in oxygen-lack conditions (Tuba & Csintalan, 1992). The flooded and aerated roots were grown quicker in comparison with roots flooded only, because of dissolved oxygen. On the bases of statistical analysis the aeration of Hoagland solution resulted in significantly longer and longer root at the moment of various experiments, while the root length increased slower in hydroponic solution without aeration. For seedlings flooded only and not aerated there was no significant difference in length of root during time period of experiment.

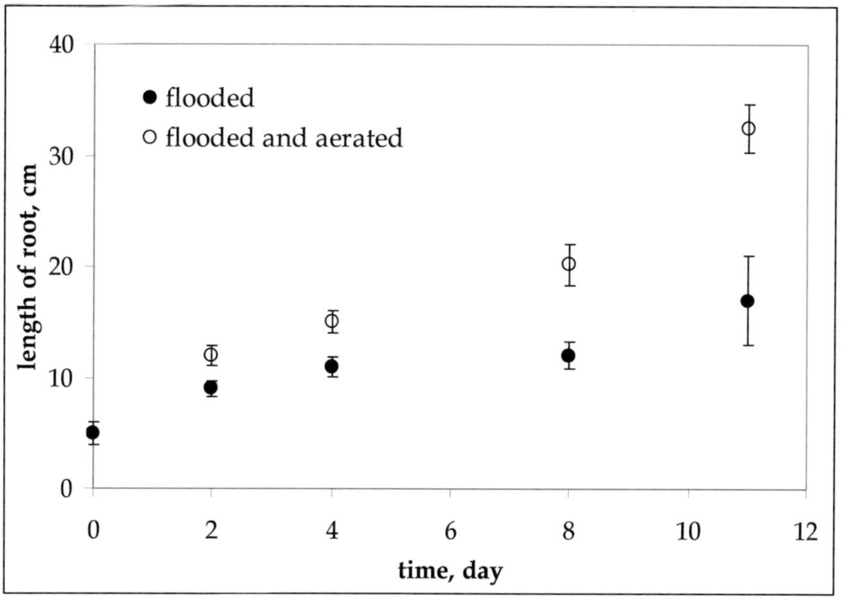

Fig. 2. The root length of three days old pea seedlings after transfer from vermiculate into hydroponic medium. Time is measured from transfer. Error bars represent SD.

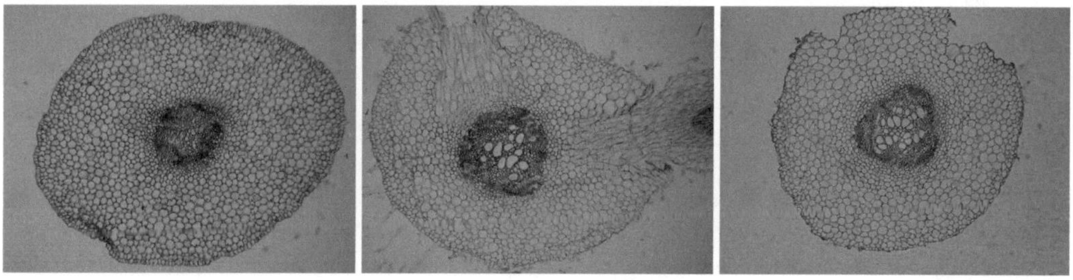

100 μm

A B C

Fig. 3. Microscopic images of cross-section of pea roots. A - Three days old pea seedling before flooding. B - Three days old pea seedling after 11 days flooding without aeration. C- Three days old pea seedlings after 11 days flooding with aeration.

The cross section images of three days old seedlings (Fig. 3A) showed the changes of the vascular cylinder of the root over time (Fig. 3B,C). The most noticeable change in the root tissue was the alteration in differentiation of tracheae of the vascular tissue in terms of numbers and size. By the end of the experiment (eleven days in Hoagland solution) the aerated controls had well developed vascular cylinder (Fig. 3C), compared to those without aeration (Fig. 3B), whose had less tracheae and we could not observe considerable amounts of small tracheae elements next to the large and developed ones.

Similar cavities were observed on scanning electron micrographs of vascular cylinder of primary pea roots grown under flooding (Sarkar et al., 2008). The vascular cavity had begun to form after 3 hours of flooding.

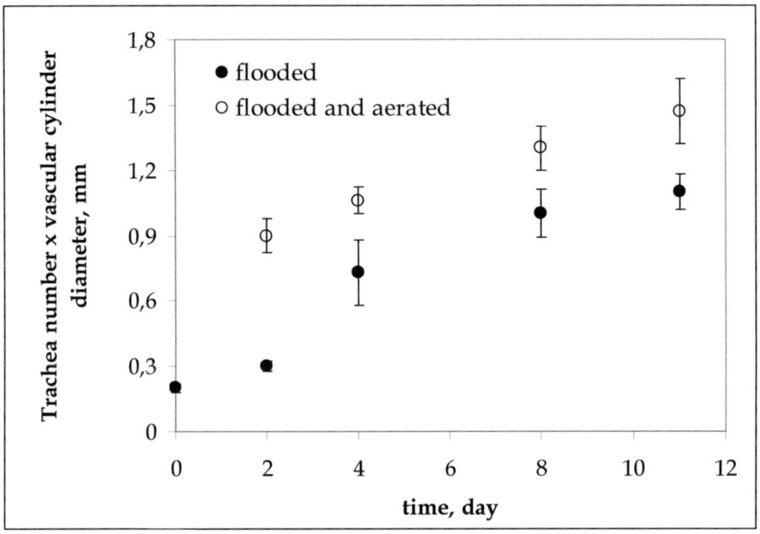

Fig. 4. Root trachea number x vascular cylinder diameter of three days old pea seedlings under flooding without aeration and under flooding with aeration in the function of flooding time. Error bars represent SD.

The state of the vascular cylinder can be characterised by the numbers of tracheae and size of the vascular cylinder, so the product of the tracheae number (T) and the diameter of the vascular cylinder (VC) was calculated as an informative parameter of the vascular cylinder (Fig. 4).

The lack of oxygen (only flooded seedlings) decreases the trachea number and vascular cylinder diameter (Fig. 4). The total area of tracheae elements of the cross sections in flooded pea seedlings with aeration was larger than in the cross sections of flooded seedlings without aeration.

3.2 Changes in the electrical impedance parameters

3.2.1 Electrical impedance spectra

The measured spectra after open and short correction were used for approaching procedure which results the impedance parameters. Typical electrical impedance spectra (in the complex plane) of the root of a seedling consist of a circular arc - at high frequencies - and a straight line - at low frequency range (Fig. 5). Similar electrical impedance curve can be taken for other biological objects, for roots of various plants (Cao et al., 2011; Rajkai et al., 2005). Generally the increase of water content in plant tissues causes the impedance decrease.

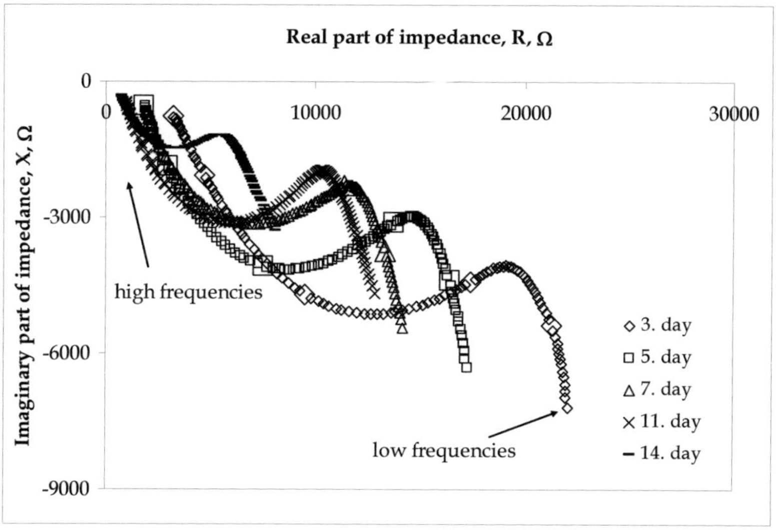

Fig. 5. Typical impedance spectra of three days old seedling at the moment of transfer from vermiculate to the half Hoagland solution and at two, four, eight and eleven days after flooding. In each spectrum the first, the second, the third, the fourth and the last enlarged point from the low frequency direction belong to frequencies of 100 Hz, 1 kHz, 10 kHz, 100 kHz and 800 kHz, respectively.

The circular arc (Fig. 5) describes the living tissue impedance, which depends on structure and state of biological object (Grimnes & Martinsen, 2000). The low frequency part (10 Hz - 1000 Hz) of the spectrum represents the electrode impedance (Macdonald, 1987). Elimination of electrode impedance is possible with a four-electrode measurement or with a correction if the impedance is measured at least with three different electrode distances (Grimnes &Martinsen, 2000; Zhang & Willison, 1991). In our experiments the diameter and length of seedlings were too little for four-electrode measurements and for more than one electrode distance. It can be assumed, that electrode polarization does not depend on the age of seedlings. In this case during our experiments the decreasing tendency in radius of Wessel-diagram can be thought truth.

3.2.2 Approaching of electrical impedance spectrum of root

For mathematical approaching the circular arc - in frequency range from 2000 Hz up to 1MHz - is used. This frequency range was used for approaching to avoid the effect of electrode impedance on parameters of tissue impedance. The measured impedance points at various frequencies were approached with modified Hayden model containing the resistance of intercellular and intracellular parts and the capacitance of cell membrane. Approaching of a typical measured spectrum with complex non-linear squares can be seen in Fig. 6. MATLAB program was used for the calculation.

The model parameters, R_a, R_s and C_m were evaluated for each seedling. The experiments were repeated three times and in one experiment the impedance spectrum of about 15-20 seedlings were measured. The average of one parameter is calculated from about 45-60 values.

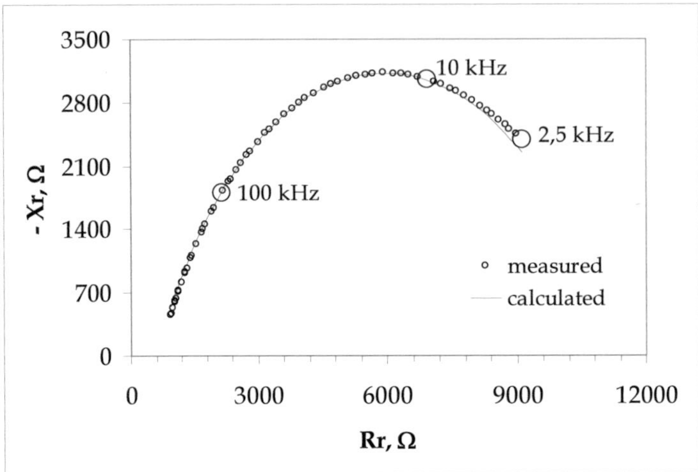

Fig. 6. Approach with complex non-linear least square method a typical measured spectrum in frequency range from 2,5 kHz up to 1 MHz. The modified Hayden model was used for approaching. MATLAB program was used for calculations.

3.2.3 Changes in electrical impedance parameters

The apoplasmic resistance, R_a, of three days old seedlings grown on vermiculate has about 25 kΩ (Fig. 7). After transferring the seedling into the half Hoagland solution this value decreases to about 10 kΩ. The total lack of oxygen (only flooded seedlings) results into a higher rate in R_a decrease. In the aerated solution this decrease is slower.

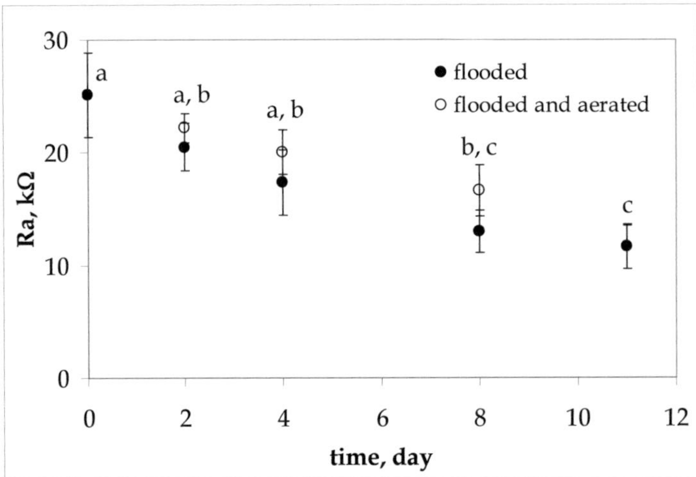

Fig. 7. Change of apoplasmic resistance (R_a) in root of three days old pea seedlings after transfer from vermiculate into half Hoagland solution with and without aeration during eleven days. The error bars represent the standard deviation. Values designated by the same letters did not differ significantly from one another in the Duncan multiple comparison test at the $p < 0.05$.

The decrease of intercellular resistance may be the consequence of interaction of several flood-induced mechanisms with some growing processes. From our light microscopic images (Fig. 3) and from scanning electron microscopic images (Sarkar et al., 2008) the tracheae formatted under flooding with higher diameters can be consider as electrical conductor. Moreover, the natural growth phenomena of pea seedlings lead also to the formation tracheae. If the cross-section of an electrical conductor increases, the resistance becomes lower.

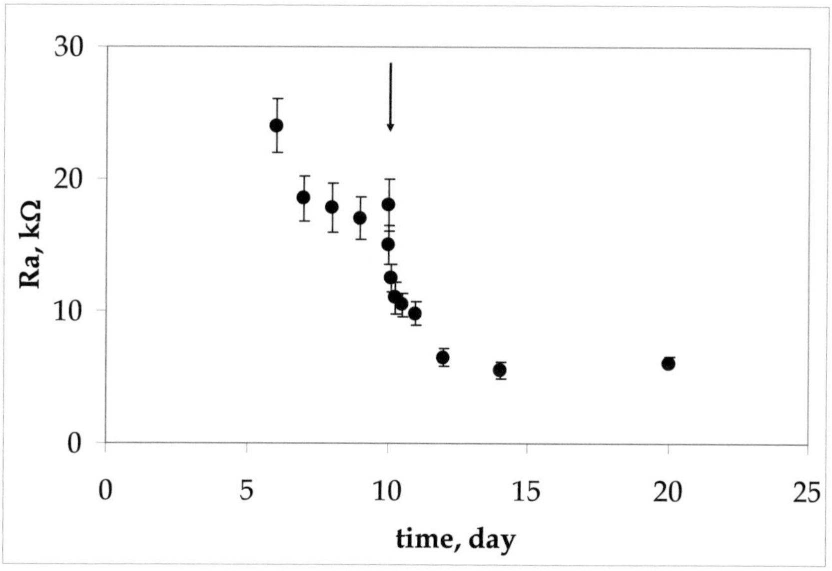

Fig. 8. Change of apoplasmic resistance (R_a) in root of ten days old seedling during growth in vermiculate and after transfer from vermiculate into half Hoagland solution without aeration. The arrow shows the transfer. The error bars represent the standard deviation.

The intercellular resistance (R_a) of ten days old seedlings after transfer from vermiculate into the hydroponic drastically decreases from about 20 kΩ down to 10 kΩ during some hours (Fig.8). This very quick decrease (during some hours) of apoplasm resistance is in good agreement of the observation, that the vascular cavity can be formed during three hours after flooding (Sarkar et al., 2008).

The resistance of symplasm in root of three days old pea seedlings also shows a decreasing tendency after flooding (Fig. 9). The R_s value decreased from about 3 kΩ down to 0.5 kΩ during our experiment time period. Practically there is no difference of intracellular resistance in root of seedlings flooded with or without aeration. But there are significant differences in R_s values of experiments following each other in time, except the two last experiments.

In the case of ten days old seedlings the intracellular resistance also drastically decreases after transfer from vermiculate to half Hoagland solution (Fig. 10). The rate of decrease is not as high as in intracellular resistance (Fig. 8). R_s value decreased from about 1,3 kΩ down to 0.5 kΩ.

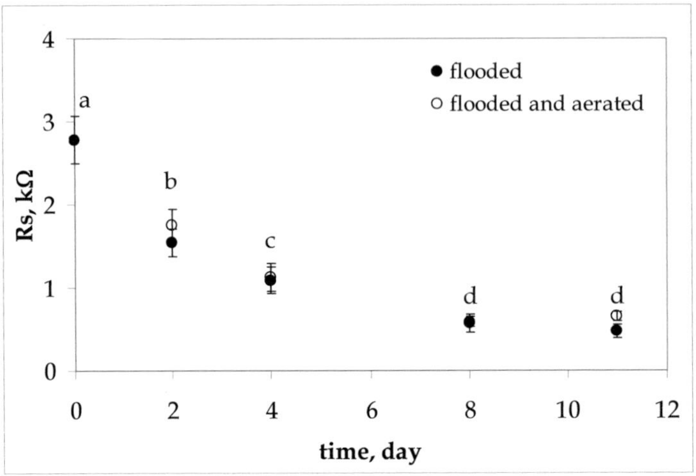

Fig. 9. Change of symplasmic resistance (R_s) in root of three days old pea seedlings after transfer from vermiculate into half Hoagland solution with and without aeration during eleven days. The error bars represent the standard deviation. Values designated by the same letters did not differ significantly from one another in the Duncan multiple comparison test at the $p < 0.05$.

Water gets into plant roots mainly via the apoplasmic route, but it can also enter cells through aquaporins - cell membrane protein channels that allow the active passage of water (Martinoia et al., 2000). The latter mechanism allows the amount of water to increase rapidly in flooded pea root tissues and this in turn increases ion mobility, which is one of the main factors affecting plant tissue impedance (Vozáry et al., 1999).

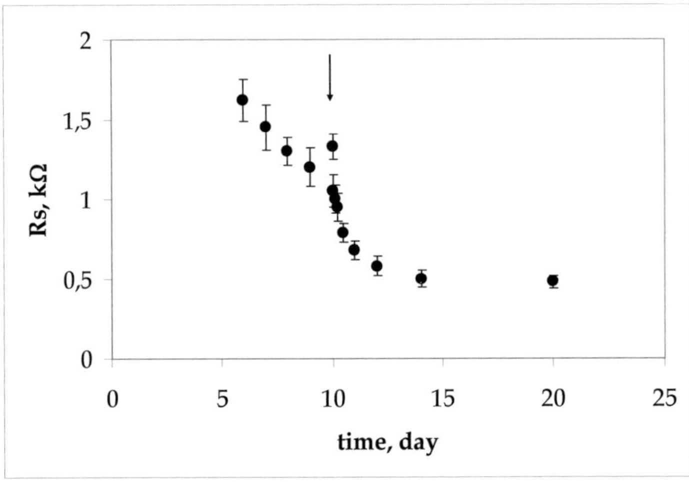

Fig. 10. Change of symplasmic resistance (R_s) in root of ten days old seedling during growth in vermiculate and after transfer from vermiculate into half Hoagland solution without aeration. The arrow shows the transfer. The error bars represent the standard deviation.

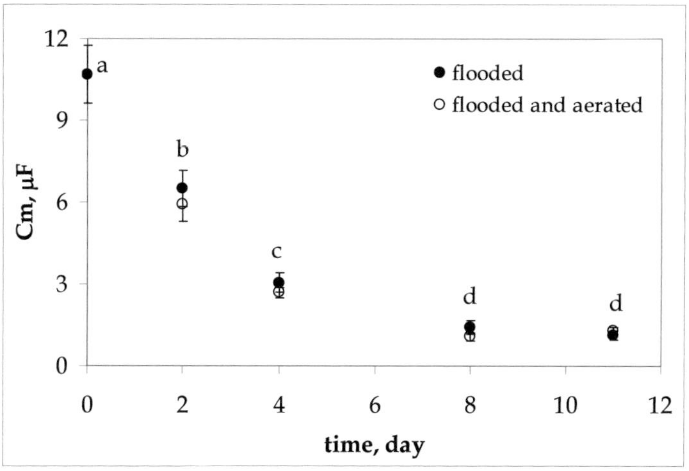

Fig. 11. Change in membrane capacitance of root in three days old pea seedlings after transfer from vermiculate to half Hoagland solution with and without aeration, during eleven days. The error bars represent the standard deviation. Values designated by the same letters did not differ significantly from one another in the Duncan multiple comparison test at the $p < 0.05$.

Sudden flooding induces the growth of vacuoles, which are also part of the electrical conductor system of the cells, and this can also contribute to decreased resistance and capacitance values (Niki & Gladish, 2001). Tissue impedance is closely related to both cellular ionic mobility (Vozáry et al., 1999) and the diameter of electrical conductors. The changes in these two factors will also lead to a decrease of symplasmic resistance and membrane capacitance values in the roots. Indeed, the decrease of membrane capacitance of root in three days old pea seedling was observed (Fig. 11) after transfer of seedlings from perlite to half Hoagland solution. The capacitance decreases from about 10 μF to 1 μF during the experiment period. The average value of membrane capacitance practically does not depend on the aeration of solutions.

4. Conclusions

Considering all these observations we conclude that parameters calculated from the measured impedance spectra constitute a non-invasive method by which it is possible to follow the changes of pea seedlings' tissue structure caused by simulated flooding. In addition, it fulfils important criteria, such as the measurement being carried out rapidly, and not requiring the processing of plant tissues. Furthermore electrical impedance measurements can be done on relatively young seedlings, so that the effects of stress agents can be detected even in the early stages of the development before visible symptoms occurred. In this developmental stage, highly sensitive environmental stress detecting methods, for example fluorescence induction measurement cannot be realised because of the lack of fully opened leaves. The method of electrical impedance measurement can be suitable for the detection of flood induced plant tissue structure changes through the alteration of R_a, R_s and C_m parameters even in the very early stage of stress evolution.

The changes in R_a, R_s and C_m parameters consist of two different phenomena: 1. the structural changes in vascular cylinder in root of pea seedlings caused by flood with or without aerations and 2. the structural changes in consequence of growing processes (increasing number and size of cells) far from central part of root. Parameter R_a seems to be sensitive to existence or lack of aeration in flooded circumstances.

Considering the possible effect of electrode polarization in consequence of overlapping the impedance spectrum of electrode with the spectrum of root tissue, the concrete values of impedance parameters can be a little changed, but the decreasing tendency of R_a, R_s and C_m after flooding has to be unchanged.

On the basic of this work the development of a portable LCR meter can be realized, which can work with some fixed frequencies and with electrodes of little geometrical dimensions and more electrode distances to eliminate the electrode polarization. This potable equipment can be used in detection of structural changes in plant tissues caused by various environmental effects, for example the extreme water supply or in contrary the drought.

5. Acknowledgment

This work was supported by Hungarian Government Foundation

TÁMOP 4.2.1. B-09/1/KMR: TÁMOP 4.2.1.B-09/1/KMR

6. References

Cao, Y.; Repo, T.; Silvennoinen, R.; Lehto, T. & Pelkonen, P. (2011). Analysis of the Willow Root System by Electrical Impedance Spectroscopy. *Journal of Experimental Botany*, Vol.62, No.1, pp. 351-358, ISSN 0022 0957

Greenham, C.G.; Randall, J.P. & Ward, M.M. (1972). Impedance parameters in relation to phosphorus and calcium deficiencies in subterranean clover (*Trifolium subterraneum* L). *Journal of Experimental Botany*, Vol.23, No.1, pp. 197-209, ISSN 0022 0957

Grichko, V.P. & Glick, B.R. (2001). Ethylene and flooding stress in plants. *Plant Physiology and Biochemistry*, Vol.39, No.1, pp. 1–9, ISSN 0981-9428

Grimnes, S. & Martinsen, O.G. (2000), Electrical properties of tissue. In: *Bioimpedance and Bioelectricity Basics*, S. Grimnes & O.G. Martinsen (Eds.), 195-239, Academic Press, ISBN 0-12-303260-1, New York, USA

Harker, F. R. & Maindonald, J. H. (1994). Ripening of Nectarine Fruit (Changes in the Cell Wall, Vacuole, and Membranes Detected Using Electrical Impedance Measurements). *Plant Physiology*, Vol.106, No.1, pp. 106-165, ISSN 0032-0889

Hayden, R.I.; Moyse, C.A.; Cadler, F.W.; Crawford, D.P. & Fensom D.S. (1969). Electrical impedance studies on potato and alfalfa tissue. *Journal of Experimental Botany*, Vol.20, No.2, pp. 177-200, ISSN 0022 0957

Hoagland, D.R. & Arnon, A.J. (1950). The water culture method for growing plants without soil. *California Agriculture Experimental Station Circular*, Vol.347, pp. 1-32,

Honda, M. (1989). *The Impedance Measurement Handbook* (A Guide to Measurement Technology and Techniques), Yokogawa-Hewlett-Packard LTD., Hewlett-Packard Co., USA

Iijima, M., & Kato, J. (2007). Combined soil physical stress of soil drying, anaerobiosis and mechanical impedance to seedling root growth of four crop species. *Plant Production Science,* Vol.10, No.4, pp. 451-459, ISSN 1343-943X

Luxova, M. (1986). The hydraulic safety zone at the base of barley roots. *Planta,* Vol.169, No.4, pp. 465-470, ISSN 0032-0935

Macdonald, J.R. (1987). *Impedance spectroscopy,* John Wiley and Sons ISBN 0-471-83122-0, New York, USA

Martinoia, E.; Massonneau, A. & Frangne, N. (2000). Transport processes of solutes across the vacuolar membrane of higher plants. *Plant Cell Physiology,* Vol.41, No.11, pp. 1175-1186, ISSN 0032-0781

Mészáros, P.; Vozáry, E. & Funk, D.B. (2005). Connection between moisture content and electrical parameters of apple slices during drying. *Progress in Agricultural Engineering Science,* Vol.1, No.1, pp. 95-121, ISSN 1786-335X

Niki, T. & Gladish, D.K. (2001). Changes in growth and structure of pea primary roots (*Pisum sativum* L. cv. Alaska) as a result of sudden flooding. *Plant Cell Physiology,* Vol.42, No.7, pp. 694-702, ISSN 0032-0781

Pething, R. & Kell, D. B. (1987). The passive electrical properties of biological systems: their significance in physiology, biophysics and biotechnology. *Physics in Medicine and Biologyl,* Vol.32, No.8, pp. 933-970, ISSN 0031-9155

Privé, J.P. & Zhang, M.I.N. (1996). Estimating cold stress in `Beautiful arcade` apple roots using electrical impedance analysis. *Hort Technology,* Vol.6, No.1, pp. 54-58, ISSN 1063-0198

Repo, T. & Zhang, M. I. N. (1993). Modeling woody plant tissues using distributed electrical cicuit. *Journal of Experimental Botany,* Vol.44, No.5, pp. 977-982, ISSN 0022 0957

Repo, T.; Zhang, G.; Ryyppo, A. & Rikala, R. (2000). The electrical impedance spectroscopy of Scot pine (*Prunus sylvestris* L.) shoots in relation to cold acclimation. *Journal of Experimental Botany,* Vol.51, No.12. pp. 2095-2107, ISSN 0022-0957

Rajkai, K.; Végh, K.R. & Nacsa, T. (2005). Electrical Capacitance of Roots in Relation to Plant Electrodes, Measuring Frequency and Root Media. *Acta Agronomica Hungarica,* Vol. 53, No.2, pp. 197-210, ISSN 0238-0161

Ruzin S.E. (1999). *Plant microtechnique and microscopy,* Oxford University Press, ISBN 0-19-508956-1, Oxford

Sarkar, P.; Niki, T. & Gladish, D.K. (2008). Changes in Cell Wall Ultrastucture Induced by Sudden Flooding at 25 °C in *Pisum sativum* (Fabaceae) Primary Roots. *American Journal of Botany,* Vol.95, No. 7, pp. 782-792 ISSN 0002-9122

Toyoda, K.; Farkas, I. & Kojima, H. (1994). Monitoring changes in material properties of agricultural products during heating and drying by impedance spectroscopical analysis. *Journal of Food Physics,* Vol.2, No.1, pp. 69-98, ISSN1416-3365

Tuba, Z. & Csintalan, Zs. (1992), The effect of pollution on the physiological process in plants. In: *Biological Indicators in Environmental Protection.* (M. Kovács; J. Podoni; Z. Tuba, & G. Turcsányi, (Eds.), pp. 169-191, Publishing House of the Hungarian Academy of Sciences, ISBN 963-05-6183-2, Budapest, Hungary

Vozáry, E.; László, P. & Zsivánovits, G. (1999), Impedance parameter characterizing apple bruise. *Annals of New York Academy of Sciences*, Vol.873, pp. 421-429, ISBN 1-57331-191-1

Vozáry, E.; Paine, D.H.; Kwiatkowski, J. & Taylor, A.G. (2007). Prediction of soybean and snap bean seed germinability by electrical impedance spectroscopy. *Seed Science and Technology*, Vol.35, No.1, pp. 48-64, ISSN 0251-0952

Vozáry, E.; Jónás, G.; Hitka, G.; Koncz, G.; Hanula-Kövér, G. & Benkő, P. (2009). Effect of air humidifying on impedance parameters of ice lettuce. *Proceedings of International Conference on Electromagnetic Wave Interaction with Water and Moist Substances*, pp. 398-404, ISBN 978-951-22-9940-9, Espoo, Finland, June 1-5, 2009

Vozáry, E. & Benkő, P. (2010). Non-destructive determination of impedance spectrum of fruit flash under the skin. *Proceedings of 14th International Conference on Electrical Bioimpedance, Journal of Physics: Conference Series*, Vol.224, 012142 , ISSN 1742-6596, Gaineswille, University of Florida, Florida, USA, April 4-8, 2010, Available from: http://iopscience.iop.org/1742-6596/224/1

Zhang, M.I.N. & Willison, J.H.M. (1991). Electrical impedance analysis in plant tissues: A double shell model. *Journal of Experimental Botany*, Vol.42, No.11, pp. 1465-1476, ISSN 0022 0957

Zhang, M.I.N. & Willison, J.H.M. (1993). Electrical impedance analysis in plant tissues: impedance measurement in leaves. *Journal of Experimental Botany*, Vol.44, No.8, pp. 1369-1375, ISSN 0022 0957